MEXICO'S COMMUNITY FOREST ENTERPRISES

DAVID BARTON BRAY

MEXICO'S COMMUNITY FOREST ENTERPRISES

Success on the Commons and the Seeds of a Good Anthropocene

THE UNIVERSITY OF
ARIZONA PRESS

TUCSON

The University of Arizona Press
www.uapress.arizona.edu

ISBN-13: 978-0-8165-4112-6 (hardcover)

Cover design by adam b. bohannon
Cover photo: Mexico Jungle Landscape by THPStock.
Typeset in Adobe Caslon Pro 10/14 and Trade Gothic Next (display)

Unless otherwise noted, all graphs are by the author.

Library of Congress Cataloging-in-Publication Data
Names: Bray, David Barton, author.
Title: Mexico's community forest enterprises : success on the commons and the seeds of a good anthro-
    pocene / David Barton Bray.
Description: Tucson : The University of Arizona Press, 2020. | Includes bibliographical references and
    index.
Identifiers: LCCN 2020011846 | ISBN 9780816541126 (hardcover)
Subjects: LCSH: Forest management—Mexico. | Community forestry—Mexico. | Community
    forests—Mexico.
Classification: LCC SD147 .B73 2020 | DDC 634.9/20972—dc23
LC record available at https://lccn.loc.gov/2020011846

Printed in the United States of America
♾ This paper meets the requirements of ANSI/NISO Z39.48-1992 (Permanence of Paper).

CONTENTS

ILLUSTRATIONS

FIGURES

MAPS

TABLES

ACKNOWLEDGMENTS

The debts have accumulated in the over thirty years that I have been around Mexican community forests, so I am sure to miss some here. I must first express my gratitude to Sergio Madrid, Luis Hernández Navarro, and Fernando Melo, who one day in October 1989 first took me up to the Sierra Norte of Oaxaca and thus changed my professional and later my personal life. Leticia Merino was my first Mexican academic colleague and colleague with the Inter-American Foundation, with whom I shared many trips to forest communities. She went on to write a substantial body of work on community forestry, much of which is cited in these pages. Salvador Anta, Francisco Chapela y Mendoza, Gonzalo Chapela y Mendoza, Yolanda Lara, Rodolfo López, Sergio Madrid, Victoria Santos, Rosa Ledesma, Marcelo Carreón, Alfonso Arguelles, Hugo Galletti, and the late Deocundo Acopa (in no particular order) have freely shared with me their deep knowledge of the sector for decades, and if they could somehow collectively write a book, it would be far more comprehensive and informed than this one.

I was happily doing research on the community forests of Quintana Roo, with no thought of ever looking beyond them, when in 2001 I and Leticia Merino were approached by Deborah Barry, then of the Ford Foundation, about doing a national-level research project on the subject. It was her vision of the significance of the phenomenon that allowed us to take on something I am not sure I would have ever thought of on my own, so thanks for that, Deborah.

Colleagues and myself were blessed with substantial funding over the years, so enormous gratitude for that to the former North-South Center of the University of Miami, the Fulbright-Hays Program for Mexico (twice!), the William and Flora Hewlett Foundation, the Ford Foundation, the Tinker Foundation, and the United States Agency for International Development. Thanks also to the former Department of Environmental Studies and the current Department of Earth and Environment at Florida International University for providing a supportive institutional foundation since 1997.

The many dozens of forest communities visited over the decades were uniformly open and welcoming (OK, with a few exceptions), and not only when I was able to support them with funding. They also tolerated me for many years afterward, when all I had to offer were questions and many publications in English, along with a few books and articles in Spanish. Thanks also to Rocío Aguilar Méndez and Guadalupe Pacheco Aquino for invaluable support in final manuscript preparation.

Thanks for reading and commenting on one or more chapters of the draft manuscript goes to Dan Klooster, Camille Antinori, Gustavo García-López, Prakash Kashwan, Salvador Anta, Rodolfo López Arzola, Joe Figel, Ben Hodgdon, Harini Nagendra, Gustavo Pérez-Verdín, Jim Robson, Francisco Chapela y Mendoza, and Yolanda Lara. They made the manuscript better, and it was surely a mistake to not have always accepted all of their suggestions. Thanks also to Allyson Carter of the University of Arizona Press for having the wisdom to see the value of the manuscript and the rest of the UA Press team for their superb support, as well as to Sonya Manes for her excellent copyediting.

Gratitude also goes to my adult children, Abigail Lydia Bray and Erik Madison Bray, who have suffered my absences to Mexico for most of their lives (and were occasionally able to share the adventure), and don't seem much the worse for it. Finally, I wish to express deep thanks to my life companion and best personal and professional advisor, Elvira Durán Medina, whom I first met in a community forest in Quintana Roo.

ABBREVIATIONS

ADCVS	Áreas Destinadas Voluntariamente a la Conservación (Voluntary Conservation Areas)
AMA	Acuerdo México-Alemania (Mexico-Germany Agreement)
ARS	Asociaciones Regionales de Silvicultores (Regional Silvicultural Associations)
ASETECO	Asesoría Administrativa a Empresas Comunales (Administrative Advising for Communal Enterprises)
ASILVITLAX	Asociación de Silvicultores del Municipio de Tlaxco (Association of Silviculturists of the Municipality of Tlaxco)
BOMACHIZA	Bosques y Maderas de Chignahuapan-Zacatlán (Forests and Timbers of Chignahuapan-Zacatlán)
CAR	corrective action request
CCIT	Centro Coordinador Indigenista de la Tarahumara (Indigenous Coordinating Center of the Tarahumara)
CCMSS	Consejo Civil Mexicano para la Silvicultura Sostenible (Mexican Civil Council for Sustainable Silviculture)
CDI	Comisión Nacional para el Desarrollo de los Pueblos Indígenas (National Commission for the Development of Indigenous Peoples)
CFE	community forest enterprises
CFO	Compañía Forestal de Oaxaca (Forest Company of Oaxaca)

CFUG	Community Forest User Groups (Nepal)
CIOAC	Central Independiente de Obreros Agrícolas y Campesinos (Independent Central of Agricultural Workers and Peasants)
COCOEFO	Coordinadora de Organizaciones y Ejidos Forestales de Oaxaca (Coordination of Forest Organizations and Ejidos of Oaxaca)
CBFM	Community-based Forest Management (Tanzania)
CONAFOR	Comisión Nacional Forestal (National Forest Commission)
CONANP	Comisión Nacional de Áreas Naturales Protegidas (National Commission of Natural Protected Areas)
CONASIL	Consejo Nacional de Silvicultores (National Council of Silvicultores)
COP	Conference of Parties
CPR	common pool resource
CORENCHI	Comité de Recursos Naturales de la Chinantla Alta (Natural Resource Committee of the Upper Chinantla)
DGDF	Dirección General de Desarrollo Forestal (General Directorate for Forest Development)
ERA	Estudios Rurales y Asesoría (Rural Studies and Advising)
FAPATUX	Fábricas de Papel de Tuxtepec (Tuxtepec Paper Factories)
FIFONAFE	Fideicomiso Fondo Nacional de Fomento Ejidal (National Trust Fund for Ejido Development)
FMP	Forest Management Programs
FONAFE	Fondo Nacional de Fomento Ejidal (National Fund for Ejido Promotion)
FOVIGRO	Forestal Vicente Guerrero (Vicente Guerrero Forest Company).
FRA	Forest Rights Act (India)
FSC	Forest Stewardship Council
GAIA	Grupo Autónoma para la Investigación Ambiental (Autonomous Group for Environmental Research)
GEA	Grupo de Estudios Ambientales (Environmental Studies Group)
GNP	gross national product
GTZ	*Gesellschaft für Technische Zusammenarbeit* (German Corporation for International Cooperation)
HGEU	Hermenegildo Galeana Ejido Union (Union de Ejidos Hermenegildo Galeana)
I/CCA	Indigenous/community conserved areas
ICICO	Integradora de Comunidades Indígenas y Campesinas de Oaxaca (Integrator of Indigenous and Peasant Communities of Oaxaca)

ICOFOS	Integradora Comunal Forestal de Oaxaca, S.A. de C.V. (Communal Forest Integrator of Oaxaca)
IDB	Inter-American Development Bank
IEEPO	Instituto Estatal de Educación Pública de Oaxaca (Institute of Public Education of Oaxaca)
IFM	Improved Forest Management
IFRI	International Forestry Resources and Institutions Program
INE	Instituto Nacional de Ecología (National Ecology Institute)
INI	Instituto Nacional Indigenista (National Indigenous Institute)
IXCAJIT	Ixtlán-Capulalpan-Xiacui-Trinidad
JFM	Joint Forest Management
LGDFS	Ley General de Desarrollo Forestal Sustentable (General Law of Sustainable Forest Development)
LGEEPA	Ley General del Equilibrio Ecológico y la Protección al Ambiente (General Law of Ecological Equilibrium and Environmental Protection)
LKS	Lesser-Known Tropical Species
MDS	Método de Desarrollo Silvícola (Silvicultural Development Method)
MFCS	Mexican Forest Certification System
MIQROO	Maderas Industrializadas de Quintana Roo (Industrialized Timbers of Quintana Roo)
MMOBI	Método Mexicano de Ordenación de Bosques Irregulares (Mexican Method for Ordering Irregular Forests)
MMOM	Método Mexicano de Ordenación de Montes (Mexican Method of Forest Ordering)
MREDD	Mexico Reducing Emissions from Deforestation and Forest Degradation Program
NAFTA	North American Free Trade Agreement
NIFP	nonindustrial private forests
NGO	nongovernmental organization
NOM	Norma Oficial Mexicana (Mexican Official Norm)
NTFP	non-timber forest products
ODRENASIJ	Organización en Defensa de los Recursos Naturales y Desarrollo Social de la Sierra de Juárez (Organization in Defense of Natural Resources and Social Development of the Sierra Juárez)
OEPFZM	Organización de Ejidos Productores Forestales de la Zona Maya (Organization of Forest Production Ejidos of the Mayan Zone)

OPD	Organismos Públicos Descentralizados (Decentralized Public Organisms)
OTC	Ordenamiento Territorial Comunitario (Community Territory Land-Use Zoning)
PES	Program for Environmental Services
PFA	permanent forest areas, or áreas forestales permanentes
PFM	Participatory Forest Management (Tanzania)
PHS	Payment for Hydrological Services
PPF	Plan Piloto Forestal (Forest Pilot Plan)
PROCYMAF	Programa de Conservación y Manejo Forestal (Forest Conservation and Management Program)
PRODEFOR	Programa para el Desarrollo Forestal (Forest Development Program)
PRODEPLAN	Programa de Apoyos para el Desarrollo de Plantaciones Forestales Comerciales (Support Program for the Development of Commercial Forest Plantations)
PROFAS	Fomento a la Autogestión Silvícola (Program for the Strengthening of Self-Management of Forestry)
PROFEPA	Procuraduría Federal de Protección Ambiental (Federal Attorney General for Environmental Protection)
PROFORMEX	Productos Forestales Mexicanos (Mexican Forest Products)
PROFORMICH	Promotora Forestal de Michoacán (Forest Promotor of Michoacán)
PROFOTARAH	Productos Forestales de la Tarahumara (Forest Products of the Tarahumara)
REDD+	Reduction of Emissions from Deforestation and Degradation, "plus" conservation, sustainable forest management, and enhancement of forest carbon stocks
RED MOCAF	Red Mexicana de Organizaciones Campesinas Forestales (Mexican Network of Peasant Forest Organizations)
RIL	Reduced Impact Logging
SAO	Servicios Ambientales de Oaxaca (Environmental Services of Oaxaca)
SECS	social-ecological-climate system
SEMARNAP	Secretaría de Medio Ambiente, Recursos Naturales y Pesca (Secretariat for the Environment, Natural Resources and Fisheries)
SEMARNAT	Secretaría de Medio Ambiente y Recursos Naturales (Secretary of the Environment and Natural Resources)

SES	social-ecological system
SEZARIC	Grupo Silvindustrial General Emiliano Zapata (General Emiliano Zapata Silvi-Industrial Group)
SICODESI	Sistema de Conservación y Desarrollo Silvícola (Conservation and Silvicultural Development System)
SICOBI	Sistema Comunitaria para el Manejo y Protección de la Biodiversidad (Community System for the Management and Protection of Biodiversity)
SPFEQR	Sociedad de Ejidos Productores Forestales de Quintana Roo (Society of Forest Production Ejidos of Quintana Roo)
SRA	Secretaría de Reforma Agraria (Secretary of Agrarian Reform)
TCO	Tierras Comunitarias de Origen (Community Origen Lands)
TIP MUEBLES	Textitlán-Ixtlán-Pueblos Mancomunados Muebles (Textitlán-Ixtlán-Pueblos Mancomunados Furniture)
UAF	Unidad de Administración Forestal (Forest Administration Unit)
UCEFO	Unión de Ejidos y Comunidades Forestales de Oaxaca (Union of Forest Ejidos and Communities of Oaxaca)
UCODEFO	Unidades de Conservación y Desarrollo Forestal (Units of Conservation and Forest Development)
UEHG	Unión de Ejidos Hermenegildo Galeana (Hermenegildo Galeana Ejido Union)
UIEF	Unidades Industriales de Explotación Forestal (Industrial Units for Forest Exploitation)
UMAFOR	Unidades de Manejo Forestal (Forest Management Units)
UNECOFAEZ	Unión de Ejidos y Comunidades Forestales Emiliano Zapata (Emiliano Zapata Union of Forest Ejidos and Communities)
UNFCCC	UN Framework Convention on Climate Change
UNOFOC	Unión de Organizaciones Forestales Comunitarias (Union of Community Forest Organizations)
UPMPF	Unidades de Producción de Materia Prima Forestal (Raw Forest Material Production Units)
UZACHI	Unión Zapoteca-Chinanteca (Zapotec-Chinantec Union)

MEXICO'S COMMUNITY FOREST ENTERPRISES

CHAPTER 1

STATE POLICY, MARKETS, COMMUNITY AND THE SEEDS OF A GOOD ANTHROPOCENE

INTRODUCTION

For decades, researchers concerned with sustainable management of forests in tropical countries have presented evidence that the road to reversing deforestation and improved stewardship is the devolution of greater control over forests to the local communities who obtain their livelihood from them. Inspired by these findings, various joint management, extractive reserve, indigenous territory, and community forestry initiatives have gained ground in developing countries, though the pace of progress has slackened in recent years (Rights and Resources Initiative 2018). In these efforts, governments and local communities share responsibility for the production of timber, non-timber forest products (NTFPs), and more recently ecosystem services under varying land tenure and institutional arrangements, all of through which local users receive substantial rights over forest resources. In contrast, conservationists who despair at the steady loss of tropical forests present contrasting evidence that the only way to stem the tide of deforestation is to place as many tracts as possible under strict protection through public protected areas as the essential cornerstone of biodiversity conservation (Andam et al. 2008, Laurance et al. 2012). Caught in the middle of this academic and policy debate, forest inhabitants of the developing world struggle to balance the extraction of precarious livelihoods from forests while responding to increasing pressures from national governments, international institutions, and their own perceptions of environmental decline to protect biodiversity, restore forests, and mitigate climate change.

The case for community forestry as a strategy for halting deforestation and improving management in multiple dimensions is made in this book using the example of Mexico. Community forestry in general has been widely associated with positive outcomes for forests (Nagendra, 2007; Ostrom and Nagendra 2006; Porter-Bolland et al. 2012; Rasolofoson et al. 2015). The World Resources Institute, the Rights and Resources Initiative, and other researchers have presented an array of evidence that strengthened community rights over forests are strongly associated with healthy forests, the avoidance of emissions from deforestation, and increases in carbon storage, and that these rights are more effective than public protected areas at providing an array of benefits for both human communities and the environment (Porter-Bolland et al. 2012; Stevens et al. 2014). Mexican community forest enterprises (CFEs) are the best evidence globally that local control of forests can result in a suite of positive consequences. CFEs are unique types of firms, community market-oriented enterprises managing a common forest property, with a focus on job generation rather than profits, but that nonetheless operate based on natural, physical, financial, human, and social capital (Antinori and Bray 2005).

Mexico presents a unique case in which much of the nation's forests were placed in the hands of communities, in successive degrees of actual control, stretching from the 1920s to the 1990s, due to the Mexican Revolution (1911–17). This "transformative decentralization" (Larson and Soto 2008) of forest management created Mexico's community-managed common property forest sector in both temperate and tropical areas at a scale and current level of maturity unmatched anywhere else in the world. It is thus a national laboratory for studying the political, economic, social, and ecological benefits of delivering forests to local communities with the top-down supply of governance institutions and the five capitals necessary to establish market-oriented CFEs. This institutional and capital supply opened spaces and options that were vigorously occupied and extended by community collective action. Over a thousand of these human communities in the forests of Mexico have made impressive strides over the last several decades toward achieving a "sweet spot" of extracting timber and generating incomes from their forests while conserving biodiversity and forest cover and mitigating climate change for future generations (Bray 1995; Bray 2010a; Bray et al. 2003; Bray Merino-Pérez, and Barry 2005a; Klooster and Masera 2000). A number of others are mired in corruption, conflict, and forest degradation and increasingly impacted by organized crime (CCMSS 2006). However, the large number of cases with substantial success in Mexico's forests show the enormous potential of having rights over forest resources devolved to communities of subsistence corn farmers with frequently supportive state policy. It gives community members the incentive to participate in collective action in order to take advantage of favorable state policy, defend their rights, organize community firms, and build resilience responding to the powerful market incentive of timber.

This book explores how the Mexican story of success on the forest commons has emerged, as well as where it is struggling or has failed. It is the story of Mexican community forestry, and the scientific evidence for its social and environmental achievements, and how, in its most successful manifestations, it became a global model for sustainable landscapes, environmental justice, and climate change mitigation in developing countries (Bray 2010a; Bray et al. 2003, Bray, Merino-Pérez, and Barry 2005a; Kashwan 2017; Klooster and Ambinakudige 2005). There are many regions in Mexico where CFEs, many of them forming clusters that dominate the regional landscape, have low-to-nonexistent deforestation or expanding forests, sustainable forest management, enhancement of carbon stocks, biodiversity conservation, and the generation of forest-based livelihoods. For decades Mexico has been conducting a de facto large-scale experiment in the design of a national social-ecological system (SES) focused on community forests. What happens when you give poor subsistence corn-farming communities rights over forests, training, organizational support, equipment, and financing—which is to say, natural, human, physical and financial capital—and augment social capital that some (but not all) already had? Do the farmers destroy the forests in the name of economic development, or do they manage them sustainably, generating current income while maintaining intergenerational value as a resource for their children? Researchers who have worked with Mexico's forest communities over three decades are now beginning to have some answers to these questions.

This book arises from my professional engagement with Mexican community forestry over a thirty-year period. From 1989 to 1997, I worked in Mexico as a foundation representative with the Inter-American Foundation (IAF), a U.S. government agency, where I first encountered CFEs in the Sierra Norte of Oaxaca. They were then only seven years removed from their origins as a protest movement against a twenty-five-year government concession to a parastatal forest company, but where the concession had also served as a school for industrial forestry (Bray 2007). For eight years with the IAF, I had the opportunity to visit, fund, and study the still-emerging community forestry sector. Since 1997, I have had the opportunity to conduct research and publish on the sector with multiple collaborators, many of them cited in these pages and one, Juan Manuel Torres-Rojo, is co-author of chapter 6. This book represents the culmination of both of those experiences and attempts to confront the empirical experience of Mexican CFEs with a cluster of relevant social science concepts: common property theory, resilience theory, approaches to the political economy of reforms in Mexico, and theories of SESs in order to establish a framework that can explain the emergence and implications of this sector. I argue that these approaches have provided important interpretations of the development of common property regimes but that all need to be extended to account for the large and entrepreneurially successful community forestry sector of Mexico. My training is as an anthropologist, but my intellectual evolution has been toward self-identification as an environmental social scientist. My approach is therefore quite interdisciplinary, and I run the risk of misunderstanding theoretical nuances of some approaches and will surely be open to criticism from scholars more deeply schooled in a particular discipline. With that disclaimer, I have recklessly forged ahead and attempted to cobble together an interdisciplinary understanding of the emergence of this most unusual example of entrepreneurial "success on the commons" (Agrawal and Chhatre 2006; McKean 1992). It is common even among professionals close to the sector to suggest that there are only a limited number of successful examples of CFEs in Mexico (Skutsch et al. 2018). But there is overwhelming evidence of a thousand or more cases of success, with success defined as surviving for decades, being financially profitable, maintaining forests and biodiversity, storing carbon, and delivering some level of benefits to the communities, despite varying degrees of internal and external pressures and problems. This book aspires to make both a substantive empirical contribution to our knowledge of Mexican community forestry and contributions to common property theory, resilience theory, and theories of the political economy of reform.

THE MEXICAN COMMUNITY FOREST SECTOR

Rural Mexico is distinguished by having a large-scale common property sector created over decades as an ongoing consequence of the Mexican Revolution

(1911–17). The Mexican Revolution created two forms of common property, *ejidos*, land grants to landless farmers, and *comunidades*, land grants recognizing territorial rights of indigenous peoples first authorized during Spanish colonial rule; both together will be referred to as "agrarian communities."[1] As of 2012, there were 31,837 agrarian communities in Mexico (29,490 ejidos and 2,347 comunidades) occupying 100.3 million hectares, or 51 percent of the national territory (Robles Berlanga 2012). Some 69.3 million hectares of the total 100.3 million in agrarian communities is specifically categorized as community common property, while the rest is assigned to individual agricultural parcels or village areas. In many communities the agricultural parcels are also under community control, with rules prohibiting sales to outsiders. The 69.3 million hectares constitute 35.7 percent of the national territory of Mexico. Thus, over one-third of the country is directly managed as a common pool resource (CPR) by local communities (Robles Berlanga 2012). As a significant component of this vast common property sector, communities own around 60 percent of the forests of Mexico (Madrid et al. 2009), the second-highest rate of community ownership in the world after Papua New Guinea (Rights and Resources Initiative 2018), highlighting the empirical and theoretical significance of the sector for the study of common property globally. Precise numbers of communities managing these forests can still be hard to come by and the numbers fluctuate from year to year, but the most reliable recent estimate is that for the 2011–13 period there were 1,621 communities with logging permits (Torres Rojo and Amador Callejas 2015a), though another source suggests higher numbers (Carillo-Anzures et al. 2017).

Common property forest governance institutions in Mexico for ejidos came from above as a means to organize landless peasants, while in comunidades the top-down institutions merged with traditional forms of governance such as the colonial-era *cargo* system, in a form of political syncretism. This enormous national-scale common property system was inspired by earlier indigenous institutions but is defined and rooted in the modern Mexican national constitution. Both ejidos and comunidades are formal organizations established in Mexican agrarian law. This formal organization includes the Assembly of all legal community members; the democratically elected

---

1. The English term "communities" will be used throughout to refer to both ejidos and comunidades unless it is important to distinguish between them. The Spanish term *comunidades* will be used in referring specifically to that type of agrarian reform unit.

governing council (Comisariado), consisting of president, secretary, and treasurer; and the Oversight Council (Consejo de Vigilancia; see chapters 2 and 5). This is the institutional and organizational platform of CFE administration, the market-oriented firms that manage the common property forests (Antinori and Bray 2005). Through the constitution and agrarian law, village territorial governance is systematically linked to higher levels of Mexican government administration in multilevel governance institutions (Cronkleton, Bray, and Medina 2011).

The size of the sector created the need for a classification scheme based on the degree of industrial vertical integration of the CFEs (Bray, Merino-Pérez, and Barry 2005b), developed by the Mexican forest agency in the 1990s. I have modified the original classification and proposed a new version (see table 1). The original classification runs from Types I to IV, with increasing integration of the industrial-processing value chain. Type I communities are "potential producers," those with commercial forests but who do not currently have or may never have had logging permits. Type II are those communities that contract the logging operation to outside logging companies, selling timber "on the stump"; labor participation and control by the community can run from no involvement to the community providing labor and direct supervision by a forest foreman (*jefe de monte*). In the 2011–13 period, 699 of the 1,621 communities with logging permits (43 percent of total) were Type II. Type III CFEs are producers of roundwood, defined by their ownership of extraction equipment, such as skidders and tractors for extracting logs from the forest and trucks to take logs to sawmills, a next stage in industrial vertical integration, and there were 738 of them (46 percent of the total). Finally, Type IV CFEs operate sawmills, a more sophisticated and complex stage in vertical integration, and there were 184 of them (11 percent of the total; Torres Rojo and Amador Callejas 2015a). A smaller but undefined number (at least a couple of dozen) of the Type IV sawmill communities have highly sophisticated multimillion-dollar operations with further value-added processing including dryers, packing crate, furniture, and/or plywood factories and commonly employ hundreds of workers from the community and surrounding areas. Here I propose a new type to classify them, a Type V. As will be discussed in chapter 6, increasing numbers of communities manage their forests exclusively for NTFPs, ecotourism, conservation, and/or ecosystem services, and these should be given a new classification, as Type VI (Hernández-Aguilar et al. 2017; see also chapter 5).

TABLE 1. PROPOSED TYPOLOGY OF MEXICAN CFES

TYPE	CRITERIA OF CLASSIFICATION	NUMBER OF CFES (2011–13) 1,621
Type I	**Potential producers.** Communities with capacity for sustainable commercial production that currently do not carry out logging. They may have logged in the past or may have never logged.	NA
Type II	**Producers who sell timber on the stump.** CFEs where logging is carried out by contractors. Communities may have no participation or may supply labor such as chainsaw operators and community supervision (jefes de monte) of the process.	699 (43%)
Type III	**Producers of roundwood.** CFEs who own and operate equipment that extracts logs from the forest in two possible stages: (a) extraction from felling site to logging roads (tractors and skidders) and (b) logging trucks that transport roundwood to sawmills owned by others.	738 (46%)
Type IV	**Producers of sawnwood.** CFEs that are vertically integrated from logging, through extraction and transport to owned and operated sawmills and who directly market their sawnwood.	184 (11 percent)
Type V	**Advanced value-added.** Communities that may have one or more of timber dryers, furniture factories, plywood factories, packing crates, molding factories, or other advanced processing. This category may also include CFEs that are one of the other types and have also diversified into NTFPs such as resin extraction, ecotourism, or bottled water.	No statistics available
Type VI	**Non-timber forest products.** Communities that do not produce timber but sell a range of extractive and nonextractive NTFPs such as pine resin, bottled water, ecotourism, and environmental services.	No statistics available

Source: Modified from PROCYMAF (2000); Torres Rojo and Amador Callejas (2015a).

There is an association between the degree of vertical integration and the size of the commercial forest; one estimate finds that the average size of the forest under management for Type II is 922 hectares, for Type III 1,533 hectares, and for Type IV 3,503 hectares (Bray et al. 2007) (the study did not separately classify Type V communities). There are many Type IV and V communities with over 10,000 hectares of commercial forest and a few communities with over 100,000 hectares. In this study, I will be focusing substantially on the Type IV and V communities, since they have been studied the most and show the potential of the sector. The Type II and III communities are less studied, with some exceptions discussed in later chapters, but many general observations will be made about them, and there is evidence that a number of them are successful as well, based on the criteria mentioned at the end of the previous section. Most Type IV and V communities and some Type IIIs provide an array of social benefits, including health insurance, old age pensions, and accidental death benefits, extremely uncommon in rural Mexico. Despite multiple and frequently observed inefficiencies and international competition, CFEs have been found to be profitable at all levels of vertical integration (Antinori 2005; Cubbage et al. 2105), in part due to substantial state support (see chapter 6). Economic benefits have not been well documented, but it is clear that CFEs provide thousands of relatively well-paying jobs, exceedingly rare in rural Mexico. The sector is characterized by substantial subsidies, with exceptions in some historical periods, since the 1970s. One recent study found that state subsidies in the 2002–10 period had a statistically significant positive impact on poverty alleviation over CFEs who did not receive subsidies. The effect was highest with subsidies that targeted human and social capacity building or human and social capital (Torres Rojo, Moreno-Sánchez, and Amador-Callejas 2019). One estimate suggests that about 6.2 million hectares are under formal government-approved management by the CFEs nationally (Torres Rojo, Moreno-Sánchez, and Mendoza-Briseño 2016). CONAFOR (2012a) estimates there are some 15 million hectares of forest with commercial potential (though the criteria for this number are not stated), and, if correct, around 40 percent of that number is currently under management. As one measure of good forest management, as of March 2019 the number of forests certified by the Forest Stewardship Council (FSC) had reached seventy-nine communities, with 1,079,720 hectares certified, thanks to significant government subsidies (see chapter 7) (FSC Certificate Database 2020). In many cases the forests are too small to make a substantial difference in local livelihoods, but nonetheless generate some forest-based income that alleviates poverty to minor degrees and provides some public goods.

Increased evidence shows that both timber producing and non-timber producing communities have placed major areas under both informal and formal protection, as a substantial contribution to the conservation of biodiversity

(Bray, Durán, and Molina-González 2012; Pazos-Almada and Bray 2018). For example, a community in Quintana Roo is conserving some 79 percent of its extensive forests (55,349 hectares) under varying degrees of protection by community rules (Dalle et al. 2006). By harvesting timber that goes into construction and furniture, CFEs also are producing and storing carbon in long-lived forest products, the "forest products pool" of carbon (Amacher, Ollikainen, and Koskela 2009). Consequently, CFEs contribute to the mitigation of and adaptation to climate change through maintaining and expanding forest cover for carbon capture, by providing local livelihoods that can reduce migration and resulting higher household carbon emissions in urban areas (Jurjonas and Seekamp 2019), and by jointly producing stored carbon and timber. The role of Mexican CFEs as a global model for confronting climate change in the forests of developing countries will be a major theme of the concluding chapter. All of the above are astonishing accomplishments in the creation of a national-scale forest SES based on communities who just two generations earlier were composed of subsistence corn farmers with rudimentary educations.

A large sector with a wide variety of outcomes, it has many cases in which corruption, elite capture, deforestation, and degradation are present. In particular in the last decade or so, the impacts of organized crime are causing serious problems for CFEs in several states, due to the failure of the Mexican state to enforce basic policing and administration of justice. In these cases, particularly concentrated in some regions, the common human political conditions of dysfunctional equilibria and violence are still the norm (Fukuyama 2011), and powerful local political bosses known in Mexico as caciques, in alliance with organized crime, are able to subvert democratic collective action around the commons (Angulo Carrera and Martínez Tapia 2008). In Chihuahua, many communities have ethnic divisions in which mestizo (peoples of mixed race) elites exclude indigenous people from management and benefits of the CFE (Pérez-Cirera and Lovett 2006). Many CFEs have myriad deficiencies as businesses in a competitive marketplace. Some issues include underharvesting, decapitalization, small production forests, inefficient sawmills, poor timber classification practices, managerial and organizational challenges, high transportation costs, and poor understanding of marketing (G. Chapela y Mendoza 2012, 2018a; Navia-Antezana, Marín-Togo, and Cumana-Navia 2018). The presence of these factors weighs upon the sector. As of 2018, observers began to refer to a "crisis" in the sector, linked to increasing impacts of organized crime, declines in state subsidies, perceptions of ongoing decay in levels of vertical integration and

CFEs ceasing to operate, only partly offset by the launching of new CFEs (G. Chapela y Mendoza 2018b; Torres Rojo and Amador Callejas 2015a). Despite these obstacles, many hundreds of CFEs have survived for many decades with oscillating fortunes and are able to resist or adapt to these debilitating pressures and continue to function and deliver benefits to their communities. Many other CFEs continue to thrive without being significantly impacted by these factors. With strengths outweighing weaknesses in general, Mexican community forestry remains in the vanguard of community forestry globally (Bray 2010a). Their relative success has major global implications for common property theory and practice, forest and rural development, policy reform, and the resilience of forest SESs to climate change mitigation and adaptation, to be discussed in the remainder of this chapter.

THE THEORETICAL FRAMEWORK

COMMON PROPERTY THEORY

In the effort to explain how this historically emergent phenomenon happened, I will attempt to integrate diverse threads of theory: common property, community, resilience, and approaches to the political economy of reform in Mexico, and, in a subsequent section, SES. My approach to common property theory will be through the work of Elinor Ostrom (1990, 2005) and her many colleagues. Ostrom's large and complex oeuvre challenges quick summation. She synthesized a theory and a set of powerful analytical tools that rescued the idea and practice of commons management from "the tragedy of the commons" to which it had been consigned by the biologist Garrett Hardin (1968). Hardin's argument cited the example of shepherds as rational actors, and thus as non-communicating individuals, pushing more livestock onto the village commons to extract the quickest possible gain before others did and before the resource collapsed from overgrazing. The benefit of each animal accrued to the owner, but the costs from range degradation were shared by all villagers. The tragic result is that "ruin is the destination toward which all men rush, each pursuing his own best interest in a society that believes in the freedom of the commons" (Hardin 1968, 1244). This proposition represented a fundamental misunderstanding of the commons, identified early on by S. V. Ciriacy-Wantrup and R. C. Bishop (1975). According to these theorists, there are four basic property rights regimes:

private property, public property, common property, and open-access resources, with the last being a form of "propertylessness" (Berkes et al. 1989; McKean 2000). Common property is not "everybody's property," as Hardin suggests, with no one having rights of exclusion. The situation described by Hardin correctly characterizes an open-access resource where no one has clear defensible property rights. However, the real-world commons studied by Ostrom and two generations of researchers reveals cooperation, not lack of communication. In their view, villagers were family and neighbors who communicated with each other as owners of a jointly held resource, not as atomized rational actors and individual owners. In this setting, community norms reinforce collective action, set boundaries to the resource and membership in the community, establish rules, monitor adherence to rules, and impose sanctions on members of the communities who break the rules.

Ostrom's intellectual inspirations for constructing an alternative to the tragedy of the commons were many, varied, and some unexpected, ranging from the libertarian economist James Buchanan and institutional economists such as Ciriacy-Wantrup, to the ecologist Ernst Mayr and passing through game theory and the seven-generation rule of the Haudenosaunee Confederation (Wall 2014). Several general exegeses of Ostrom's work exist (Aligica 2014; Aligica and Boettke 2009; Tarko 2017; Wall 2014), and I will not attempt another one here. Instead, I will undertake a selective review of Ostrom and her colleagues with an eye to the relevance for Mexican community forestry. In this chapter I will attempt an overview of aspects of common property theory, and in subsequent chapters I will bring in additional pertinent aspects of Ostrom's thinking.

D. Wall (2014) argues that Alexis de Tocqueville's ([1835] 2000) depiction of the highly participatory democracy of New England townships in the first decades of the nineteenth century was the single most influential source for Ostrom's argument that commons work best when they are controlled by the commoners who are using them. The roots of Ostrom's consistent focus on "self-governance" is found in the comment by her colleague and husband, Vincent Ostrom, that "I consider Tocqueville to be correct in this presupposition that democratic societies are self-governing societies, not State-governed societies" (qtd. in Wall 2014, 56), a position that led both of them to be adopted by some libertarian thinkers (Lemke 2019). These self-governed societies would also ideally develop their own constitutions, since "a constitution constructed by citizens was vital to provide a set of rules so that different individuals could cooperate" (Wall 2014, 68), though the possibility of a constitution imposed from above is not ruled out.

Much of Ostrom's synthesis flowed from the Tocquevillian vision of highly democratic and participatory local polities. However, a notable exception is her earliest study of the complex multiscale governance of aquifers in Southern California, where she recognized the role of "facilitative political regimes" and "institutional supply" (Ostrom 1990, 137, 187), both key concepts in understanding the origins of Mexican CFEs. Pumping races by local water departments were depleting the aquifer, a condition that resulted in a long, costly process where groundwater pumpers began encouraging institutions such as voluntary associations to communicate about the state of the aquifer and to coordinate actions to achieve "incremental, sequential, and self-transforming institutional change in facilitative political regimes" (Ostrom 2005, 137). The state of California supplied a court system, financial subsidies to settle disputes, and technical assistance. Supplying institutions from above reduced transaction costs for participants, and supplying them from below, through the voluntary associations, created a sustainable resource regime. With some dramatically different particulars, I will argue that institutions and organizations from both above and below created the Mexican community forest SES. In this case, facilitative support by the state—in supplying governance institutions, harvest rules, and capital from above—was first launched by both breaking up large private forest estates and redistributing state lands. These lands were transferred to local communities and started a decades-long process that converted subsistence corn farmers into owners of market-oriented CFEs with an incentive for collective action based on valuable timber.

Common property theory tends to assign a background role to the state as a facilitative regime because of the focus on methodological individualism and the rational actor model. Therefore, rational actors in "user groups," evaluating costs and benefits, decide to cooperate around commons management in a process of self-organization and self-governance (Ostrom 1990, 2005). The rational actors around "common-pool resources" (CPRs) (a term that includes both mobile resources, e.g., fisheries, and fixed common property territories, e.g., forests) understand that they require management because they are "subtractable" (the use by one reduces the capacity of others to use it) and have problems of "excludability" (it is costly to patrol boundaries and exclude others from the resource) (Berkes et al. 1989). Institutions that can confront subtractability and maintain excludability are defined as "rules" that are "shared understandings by participants about enforced prescriptions concerning what actions (or outcomes) are *required, prohibited* or *permitted*" (Ostrom 2005, 18, italics in original). "Rules-in-use" is a term she developed that covered both formal (de jure) legal rules or laws and informal (de facto) rules that were actually known and generally observed. If they were "widely ignored or unknown" then they were not rules-in-use (Tucker and Ostrom 2005, 82).

Drilling down further into the concept of institutions as rules, Ostrom modified a hierarchy of decision systems first proposed by Ciriacy-Wantrup and Bishop (1975) that governed rights over CPRs: constitutional rules, collective rules, and operational rules. Constitutional-choice rules govern collective choice (including formulation, governance, adjudication, and modification, or rules for changing rules) (Ostrom 1990). The example of a constitutional choice arena is the creation of a marketing cooperative by fisherfolk (Schlager and Ostrom 1992), an example at a far more local scale than the rooting of common property governance in the actual Mexican Constitution, as in the case of CFEs. Collective-choice rules refer to policy making for management, adjudication, or resolution of conflict that are developed by "appropriators, their officials or external authorities in making policies" (Ostrom 1990, 52) that affect the operational rules. Finally, operational rules are those that directly affect day-to-day decisions, including "appropriation" or harvesting, "provisioning" or maintenance of the harvested resource, monitoring of rules developed by the users, and sanctions for those who break the rules. In Mexico, as we shall see, most constitutional and collective choice rules and many operational rules for forest commons governance are not chosen by the communities but come from "external authorities." Accordingly, the transaction costs of self-organization and self-governance of the CFE are substantially reduced by the rules from above.

The hierarchy of rules, then, helps to define the rights over resources. These rights are categorized into rights of access, withdrawal, management, exclusion, and alienation (Schlager and Ostrom 1992). Ostrom's hierarchy unpacked the standard distinction between private, public, and common property into a continuum of a separable "bundle of rights" (Alchian and Demsetz 1972). On that continuum, one of these bundles of rights defines a common property as a legally recognized resource with a clear set of owners who can set rules around access, withdrawal, management, exclusion, and alienation and can be considered a form of jointly held private property, such as a corporation (McKean 2000). A hallmark of Ostrom's analytical framework was the condensation of rules and rights into the eight "design principles" (1990), or institutions that are associated with success on the commons. The original design principles are widely cited, but what is less cited is that she modified them with additional nuances fifteen years later (Ostrom 2005, 259), and it is these modified design principles I will use here. The modified design principles, in summary form, follow:

1. Clearly defined boundaries for both the resource system and the individuals or households who have rights to harvest. Appropriators are also "able to effectively defend the resource from outsiders" (Morrow and Hull 1996, 1643, cited in Ostrom 2005, 262).
2. Proportional equivalence between benefits and costs: rules on resource products allocated to users are related to local conditions and requirements for labor, materials, and/or monetary inputs.
3. Collective choice arrangements: individuals affected by harvesting and protection rules are included in groups that can modify rules.
4. Monitoring: monitors of biophysical conditions and user behavior are at least partially accountable to users or are users themselves.
5. Graduated sanctions: depending on seriousness and context of offense, users who violate rules-in-use are likely to receive graduated sanctions from other users and/or officials accountable to users.
6. Conflict-resolution mechanisms: users and officials have rapid access to low-cost local arenas to resolve conflict.
7. Minimal recognition of rights to organize: users have the right to devise their own institutions not challenged by external government authorities, and users have long-term tenure rights to the resource.
8. "For resources that are parts of larger systems" nested enterprises: Appropriation, provision, monitoring, enforcement, conflict resolution, and governance activities organized in multiple layers (Ostrom 2005, 259).

Ostrom also proposes what could constitute an expansion of rule number 7, that a coping method for dealing with threats to sustainability include "the creation of associations of community-governed entities" (2005, 279). The design principles have some degree of relevance for Mexican CFEs, but in important ways they do not, and a further analysis of the fit of the design principles to the Mexican case will be conducted in the concluding chapter. Of importance now is that Mexican forest communities have little influence over rules around the harvest and protection in the production areas, but they do around conservation in the entire community territory and in the organization of the CFEs, as well as in multiple aspects of community life (J. P. Robson, personal communication, January 25, 2019). Nonetheless, the state historically went well beyond a passive, noninterfering stance, instead playing an active role that is better captured in the notion of a facilitative political regime, and market incentives are notably absent as a design principle (see following discussion).

Ostrom's constant focus on local self-governance arenas led her to initially define organizations as a "strategy" (1990, 39), collective action but not an institution itself. In Mexico, the crucial importance of formal organizations in shaping the trajectory of collective action around the forest commons suggests a reliance on D. C. North's distinction that "if institutions are the rules of the game, organizations and their entrepreneurs are the players." Organizations, with their own rules, emerge due to opportunities provided by institutions; "that is, if the institutional framework rewards piracy then piratical organizations will come into existence; and if the institutional framework rewards productive activities then organizations—firms—will come into existence to engage in productive activities" (North 1993). Ostrom later accepted this distinction between institutions and organizations (Ostrom 2005, 179). Following Ostrom and North, this book is an extended examination of how state policy, the institutional framework of Mexican forests, provided much of the hierarchy of rules around forest and harvest governance and defined many of the rights that allowed for access to markets. State policy further provided crucial supplies of capitals that enabled communities to access market incentives for community collective action around CFEs (Antinori and Bray 2005). This imposition from above reduced institutional choice, but also reduced the transaction costs in creating governance rules. Rules and organizations from above, in a highly facilitative political regime, did not crowd out collective action around forest governance (Ostrom 2005) because at the same time there was a significant market incentive in relatively high and stable prices for timber.

I now turn to the role of actual human communities, not just the "user groups" basic to Ostrom's methodological individualism. Despite the use of the rational actor model, Ostrom was most inspired by New England village government, as noted above, and had an intense appreciation of the role of norms in actual communities in supporting collective action around the commons (see chapters 4 and 5). As noted earlier, in direct management of the forest most rules come from above, but, crucially, it is in the rules and organizations of the CFE where community rule-making autonomy and the development of new norms around enterprise administration had freer rein (Antinori and Rausser 2007). State policy created the institutional framework, in North's terms, for the emergence of self-governed CFEs to resist abuses and vigorously occupy and extend the governance and management spaces created, resulting in an innovative model for market-oriented commons governance.

Thus, common property theory needs to be extended in three ways to help explain the Mexican case. First, the historic role of state policy, including conflicts around the use of state power, helped create the enabling institutions for territorial governance, for the platform for community firms, and for clear rules around forest harvests and forest management. Second, markets and price play a significant role in providing economic incentives for collective action in market-oriented firms that overcome the crowding-out effects of rules imposed from above. The fact that they are firms also requires an examination of the role of the "five capitals"—financial, physical, human, social, and natural—and the state's role in providing them. Third, different preexisting communities, ranging from family networks without a territory to indigenous peoples with millennial ties to a territory, played a significant role. Access to valuable forest resources provide incentives for both disorganized communities with low social capital to rapidly organize themselves and "preorganized" indigenous communities to deepen their social capital around a territory and the organization of a CFE. It is basic to common property theory that "the characteristics of resources and social interaction in many subsistence societies present favorable conditions for the evolution of effective self-governing resource institutions" (Dietz et al. 2003, 1908). However, this focus on "subsistence societies" and "self-governing resource institutions" cannot explain Mexican community forestry and the crucial role of large-scale government support in providing institutions and organizational models, part or all of the five capitals that enterprises need to operate, and the role of markets in further driving collective action and the creation of new institutions around self-governed community enterprises. Critiques of common property theory are common (Agrawal 2001) and have recently been organized by some around the notion of "critical institutionalism" (Cleaver and de Koning 2015; García-López 2018), but this book is the first effort to critically confront common property theory with the largest and most commercially successful common property regime that exists globally.

EXTENDING THE COMMON PROPERTY MODEL TO ACCOUNT FOR MEXICAN CFES

STATE POLICY AND POLITICAL POWER

The subtitle of Ostrom's 1990 book *Governing the Commons* is "the evolution of institutions for collective action" or local institutional supply (2005, 40). She identified local institutional supply as a third way of collective action, between theories of the firm and theories of the state in managing CPRs. In firms as collective action, an entrepreneur establishes contracts with agents who give up discretion over many choices; the entrepreneur then monitors the agents' performance and can terminate contracts. Ostrom's characterization of the state, curiously, is an autocratic one, where the ruler can use coercion and sanctions to produce collective action for benefits both for the ruler and for the ruled (1990). However, in her seventh design principle she suggested that the primary role of a more democratic state is primarily passive, to not impede self-organization and self-governance. In this view, the theory holds that collective action around the commons is optimally organized *outside* of firms and states. However, Mexico presents a dilemma for this view. In Mexico we find the phenomenon of entire communities acting as entrepreneurs in organizing firms to manage a common property, or "the community as entrepreneurial firm" (Antinori and Bray 2005). As well, the state, far from being a passive nonintrusive presence, supplied most of the institutions and organizations for governance of the forests, going much further in institutional supply than the groundwater management regime in Southern California. Agrawal (2001, 1656) has noted the absence of an important role of the state in much of common property theory: "As the ultimate guarantor of property rights arrangements, the role of the state and over-arching governance structures is perhaps central in the functioning of common pool resources." Ostrom was somewhat ambiguous about the role of the state in local common property regimes in the developing world, not always transferring the Southern California lessons. Her notion of polycentricity—that is, multiple coordinated governance units, equated with multilevel governance—was suspicious of overreliance on large central governments (Ostrom and Janssen 2004). In her development of a formal SES framework (McGinnis and Ostrom 2014; Ostrom 2007, 2009), government policy plays a limited role; rather, the focus of her model is on variables internal to the local systems (see the section "The

Social-Ecological System of Mexican Community Forestry"). She did recognize that the "absence of supportive, large-scale institutional arrangements may be just as much a threat to long-term sustenance as the presence of preemptive large-scale government agencies" (Ostrom 2005, 278), but the example she gave is of the U.S. Geological Survey providing scientific research for groundwater management, suggesting a still limited view of the role of the state. On the other hand, and in addition to the notion of a facilitative political regime, she also stressed the importance of "the characteristics of external political regimes" on collective action in rule-making (Ostrom 1990, 91). Along these lines, she noted the importance of long-term, complementary roles for government in providing institutions from above in support of local rule-making in the examples of Maine lobster and Pacific salmon fisheries (Acheson 2003; Ostrom 2005). She also followed Alcorn and Toledo (1998) in recognizing the role of supportive institutions at the national level in Mexico (Ostrom 2005, 284). Nonetheless, the present book significantly expands Ostrom's notion of what can be accomplished by facilitative political regimes supplying institutions and organizations in the case of Mexican CFEs.

One of the major lessons of institutional economics is that institutions "are created and persist because they are useful mechanisms for economizing upon transaction costs" (Tarko 2017, 7). The supply of institutions for the creation of the Mexican CFE sector greatly reduced transaction costs. In addition to supplying institutions and organizations for forest territorial governance, the Mexican state also supplied models for the organization of CFEs and crucial supplies of the five capitals necessary for enterprise success. Further, Mexican rule-making occupied a space frequently considered essential to success on the commons: rules around the harvests. "If a group of users is going to harvest from a resource over the long run they must devise rules related to how much, when and how different products are to be harvested" (Ostrom 2005, 262). As will be discussed further in chapter 5, basically all of the harvest rules for both timber and non-timber forest products are imposed from above. Much of common property theory is an argument against "the view that template forest policies are likely to work when imposed on a country as a whole" (Gibson, McKean, and Ostrom 2000, 5), but template forest policies were precisely imposed from above in Mexico. Many of the rules aimed at biodiversity conservation and forest protection in the production areas are also imposed from above, but it is in conservation of the rest of the territory that Mexican forest communities have substantial autonomy. This effort goes "beyond harvests in the commons" (Bray,

Durán and Molina-González 2012) to include forest protection, biodiversity conservation, and climate change (Moran and Ostrom 2005; Ostrom 2010).

MARKETS AND COMMON PROPERTY THEORY

Ciriacy-Wantrup and Bishop (1975, 718) posed the question, "Can common ownership of resources perform well in a market economy?" and suggest that the continued existence of grazing on pasture commons by villagers in the European Alps argues that it can. However, the general answer has been that it cannot. Historically, going back to the enclosure of the commons in eighteenth-century England due to rising market prices for agricultural production, markets have undermined the commons (Arnold 1998). A. Agrawal has noted the wide "agreement that the increasing integration with markets usually has an adverse impact on the management of common pool resources" (2001, 1656), with results such as rising inequality and declining forest conditions (Tucker 1999; Tucker, Randolph, and Castellanos 2007). Successful commons management has been explicitly placed outside of markets, since in the case of CPR regimes "CPR situations are rarely as powerful in driving participants . . . toward efficiency as are competitive markets. Nor is there any single variable, such as market price, that can be used as the foundation for making rational choices in a CPR environment. Simply following short term profit maximization in response to the market price for a resource unit may, in a CPR environment, be exactly the strategy that will destroy the CPR" (Ostrom 1990, 207). Thus, the role of economic incentives is underdeveloped in common property theory, and it appears to be mostly focused on what could be called subsistence or livelihood incentives, not market prices. Ostrom did recognize that "market conditions for resource units" could be an important situational variable for institutional choice (197), but in general incentives are structured by institutions and are perceived within a "set of working rules, combined with the relevant individual physical and social variables" (24), not suggesting a strong role for market incentives.

A subordinate role for price incentives is also evident in the conceptualization of SES (Ostrom 2009). "Economic value" is a second-level variable and not considered as one of the ten most important characteristics for successful commons management. Likewise, "market incentives" are only one of six possible "settings," with "economic development" being another. However, these settings appear to be relatively passive; the focus is on the internal dynamics of the system. Of thirty-three variables identified as important influences on four core subsystems, only two are economic, and in the category of eight "interactions," only one, "investments," is market related. "Outcomes" does not include economic or policy outcomes. As well, in Ostrom's last major book-length theoretical statement (2005), "markets" does not appear in the index.

However, Ostrom did recognize the significance of large-scale commercial fisheries driven by market incentives, as in the case of Maine lobster fisheries and Pacific salmon fisheries (Ostrom 2005, 284–86; Schlager and Ostrom 1992), though she did not systematically incorporate them into her theory. Maine lobster fishermen are individual appropriators, negotiating some of their own rules in spaces provided by legislation from the state of Maine. The Maine lobster fisheries are considered to be highly successful, but conflicts over harvest or withdrawal rules (trap limits), access rules (limited entry to new fishermen), and boundary problems have created constant tensions (Acheson 2003). It is apparently high market prices for lobster and the "very high incomes" (Acheson 2003, 128) to be earned in the industry that have encouraged the fisherman to continue working through these issues through collective action. Based on these observations, common property theory needs to be also extended to more explicitly include empirical situations such as Maine lobster fisheries and Mexican CFEs, market-based institutions in which price and profits are highly relevant. In these cases, market integration does not result in destruction of the commons but can actually strengthen collective action. As well, as firms operating in a capitalist economy, commons-based enterprises need natural, physical, human, social, and financial capital.

THE FIVE CAPITALS

Capital is defined as "stocks that have the capacity to produce flows of economically desirable outputs" (Goodwin 2003, 1). Common property theory has concentrated almost exclusively on the role of social capital in successful commons management (Adger 2000; Ostrom and Ahn 2003; Pretty 2003). However, firms require more than just social capital, and a "hypothetical model of sustainable capitalism" proposes that the five essential capitals are natural, human, social, manufactured or physical, and financial (Porritt 2007, 138). J. Porritt (2007) suggests that the two primordial capitals necessary for production were natural and human, human intelligence applied to the physical environment. Given the increasing recognition of the role of cooperation in human evolution (Bowles and Gintis 2011), social capital should be added to these primordial capitals.

Some recent formulations of the five capitals concept ignore financial capital, displaying a striking obliviousness to how contemporary economies can achieve sustainability (Costanza et al. 2014; Matson, Clark, and Andersson 2016). However, Porritt (2007, 140) argues that the five capitals, including financial "judiciously combined by entrepreneurs, are the essential ingredients of modern industrial productivity" and that "sustainability depends upon maintaining and where possible, increasing stocks of certain kinds of capital so that we learn to live off the flows (the 'income') without depleting the stock of capital itself" (143). Although Mexican CFEs are a very unusual form of firm, as we shall review in chapter 6, their possibilities for sustainability and industrial productivity depend upon judicious entrepreneurial combinations of the five capitals.

Financial capital is money used to produce something (Goodwin 2003). The source of the money can include private investment, government subsidies, low-interest loans, or reinvestment of profits into the firm. In Mexican CFEs, the most common source of financial capital has been government subsidies and reinvestment of profits into the enterprise, and in Durango in the 1960s, as one example, from private and parastatal enterprises. *Produced or manufactured* capital is "material goods that contribute to the production process" but are not part of production output (Porritt 2007, 183). In the case of forestry, this capital may include logging roads, chainsaws, logging trucks, and sawmills. *Human capital* refers to levels of education or technical skills that individuals in an enterprise have or may acquire. This may be formal schooling, but in Mexican CFEs it more commonly occurs as short-term training courses and on-the-job training in specific skills from tree felling to sawmill operations to accounting. However, in recent years, and with decades of CFE household incomes invested in education, increasing numbers of CFE foresters and administrators are community members with professional degrees. *Natural capital* can be any expression of nature used in a productive process. For instance, forests that are logged for their timber are natural capital. If the forests are not managed using sustained yield and broader sustainability practices, then degradation of natural capital can occur. Forests as natural capital can be divided into a renewable resources or environmental goods (if well managed) and environmental services (hydrological services, carbon sequestration), with the latter not being properly valued in the marketplace. *Social capital* is harder to quantify as a stock than the others, and has been defined as "the stock of trust, mutual understanding, shared values, and socially held knowledge that facilitates the social coordination of economic activity" (Goodwin 2003, 6). Institutions and organizations are essential forms of social capital, along with community norms. In the case of Mexican CFEs, the governance platform provided by agrarian laws and forest and environmental laws reduced transaction costs for social capital by creating the framework for the generation of the institutional and organizational innovations of CFEs, augmented by community norms.

THE ROLE OF COMMUNITY

The theoretical power of Ostrom's ideas comes from their basis in the rational actor model, why it is rational for individuals to cooperate around management of the commons. However, her emphasis on the crucial role of shared norms meant that the real-world examples were from communities where "norms of behavior . . . affect the way alternatives are perceived" (Ostrom 1990, 35) or what R. Axelrod (1984) called the "large shadow" of highly probable future interactions that condition behavior. As Francis Fukuyama (2014, 8) notes, "Human beings by nature are . . . norm-creating and norm-following creatures. They create rules for themselves that regulate social interaction and make possible the collective action of groups." In this vein, people do not just have incentives and create rules, but also have "motivations arising from the interpenetration of the self with culturally embedded and culture-embedding institutions" or a "thicker view of the institutions governing the commons" (Klooster 2000, 17). It is tempting to ascribe the success of many Mexican forest communities to preexisting norms of indigenous solidarity or to cultural expectations of cooperation. In Mexico, many indigenous communities do indeed have strong prosocial norms or "multilateral enforcement of group norms" (Bowles and Gintis 1998), which greatly facilitated the quick adaptation to official governance institutions and harvest rules. However, solidarity is not always the case. Daniel Klooster (2000, 17) notes that there are frequently struggles over decisions around the commons and that communities are sites of "contestation, creation, and maintenance—not only of rules—but also of the social norms that motivate an individual's action in the commons." There are many examples from Oaxaca to Chihuahua of communities in which ethnic differences and geographical dispersion of communities leads to a concentration of power and corruption by community elites (Klooster 2000; Pérez-Cirera and Lovett 2006). There is also a long history of community conflicts over management of the income from the forest (see chapter 5; see also Guerra Lizárraga 1991). The range of actual experiences in Mexico makes the assumption of preexisting prosocial norms presumptuous. There are many successful examples of CFEs with roots in indigenous culture, but there are also successful examples of communities who have norms of cooperation that evolved from a historical foundation of extreme disorganization and violence or of communities established by colonists from elsewhere that have only existed a few decades (DiGiano, Ellis, and Keys 2013; Torres Rojo, Guevara-Sanginés, and D. B. Bray 2005; Wexler 1995; see also chapter 4). The identification of either ancient or newly formed communities with a territory over which they have secure property rights appears to be crucial (see chapters 4 and 5). Therefore, the combination of reduced transaction costs from top-down institutional supply and strong market incentives acting upon a community with a territory can further build the stock of social capital or actually create it where little existed before. The extension of common property theory to include state policies, market incentives, the five capitals, and the role of communities with rights over a territory are the components of policy design that point the way to the future of forest resilience in the face of climate change.

RESILIENCE AND MEXICAN CFES

The institutional evolution of Mexican forests has produced a sector with hundreds of examples of CFEs that appear to be resilient to multiple shocks or political, social, economic, and ecological "disturbances" and with notable capacity to mitigate and adapt to climate change. Resilience theory has limitations in explaining this success. Resilience is a highly problematic concept when applied to ecological systems and becomes even more so in the effort to apply it to social-ecological systems. Social-ecological systems are virtually always resilient by some measure; we humans respond to crises by adapting as best we can. Yet the concept of resilience is so heuristically powerful, it remains both unavoidable and is an important component of notions of sustainability (Anderies et al. 2013). However, the apparent resilience and sustainability of significant portions of the Mexican community forestry SES are based on multiple dimensions not usually captured by resilience theory. The SES is politically, economically, socially, and ecologically resilient and contributes to mitigation of and adaptation to climate change. Few of these dimensions of Mexican CFEs are discussed in most of resilience theory. Proposals for enhancing forest resilience to climate change mostly focus on technical decisions that managers should make (Messier et al. 2015; Millar, Stephenson, and Stephens 2007). Efforts to conceptualize the idea of a resilient community forest SES are limited to ideas of social infra-structure, social learning, and social capital (Adger 2000; Berkes and Ross 2013). C. Folke proposed two ways to interpret resilience in an SES: (1) "adaptability," the capacity to absorb disturbances and still maintain crucial functions, and (2) "transformability," the capacity for renewal, reorganization, and development in response to disturbances that transform the system (Folke 2006, 263; Walker et al. 2004). The former suggests a dynamic stable state within a given "basin of attraction," whereas the latter suggests a transition to a different alternative stable state that may be more optimal. These are useful heuristics for Mexican community forestry, which can be conceptualized as having passed through a transformation from a stable state of state-directed forestry to a stable state, a different basin of attraction, of community-directed community forestry, which is well termed a "fundamentally new social-ecological system" (Folke 2006, 263). The sector's persistent existence suggests adaptability to political, economic, ecological, and climate disturbances under discussion throughout this book. However, as will be addressed in subsequent chapters, and summarized in the concluding chapter, resilience theory also needs to be expanded to explain the resilience of Mexican CFEs, how the resilient CFE system emerged, and its implications for climate change mitigation and adaptation.

THE POLITICAL ECONOMY OF REFORM IN MEXICO

In Mexico, there were two seismic political economic shifts that opened spaces for the emergence of CFEs. The first one was the political earthquake of the Mexican Revolution (1911–17), the kind of structural upheaval little considered in rational actor common property theory (Wall 2014). The Mexican Revolution was in part a reaction to the domination of the Mexican economy by foreign capital during the reign of the dictator Porfirio Díaz (1876–80, 1884–1911; Tardanico 1980), including timber. Timber has been one of the historic "commodity frontiers" driven by the global accumulation of capital (J. Williams 2007), and foreign capital penetrated Mexican forests in Chihuahua and Durango early in the twentieth century (Boyer 2015). However, the Mexican Revolution mostly closed off foreign capital investments in timber and other sectors of the economy, and logging became dominated by national enterprises, usually family companies with limited capital (Boyer 2015; Tardanico 1980). The modern Mexican political economy emerged as a state-directed economy in the Lázaro

Cárdenas period (1934–40) and "constituted structural changes with long-term implications for Mexico's social and economic development" (Fox 1992; N. Hamilton 1982, 274). One of these little-recognized structural changes was the beginnings of the creation of a local political-economic and institutional space, insulated from large-scale capital, where community forest entrepreneurial activity could take place and build resilience. The process of setting up a governance template for all Mexican agrarian communities, including forest communities, created a structure of constitutional, collective, and operational choice rules. The legal framework also defined most, but not all, of the rules of access, withdrawal, management, exclusion, and alienation; by doing so, it substantially reduced community transaction costs in determining these rules.

Cárdenas called for agrarian communities to function as the most important rural suppliers of food and raw materials (Fox 1992), and one essential raw material was timber. Beginning in the 1930s the agrarian reform process handed out enormous tracts of forest natural capital to communities in Durango (Hernández Astorga 2015) and elsewhere. As we will see in chapter 2, the formation of 866 forestry cooperatives by 1940 (Boyer 2015) was a product of this strategy. Most of these cooperatives failed in the short term, but in a number of cases forest production slowly evolved over the decades to come under community control. The cooperatives were also an initial expression of the revolutionary ideal that communities should control their own forest resources. As the agrarian reform continued in subsequent decades, advancing into mountainous areas and the tropics, more and more of the forests' natural capital ended up being redistributed, as in Durango and Chihuahua in the 1950s and 1960s, and in Guerrero and Chihuahua in the 1970s. These advances were politically possible because, as noted, national forest capital was mostly small and local and with reduced political power.

The second great seismic shift in the Mexican rural political economy and policy making was Luis Echeverría's presidency (1970–76). Sweeping agrarian reforms by President Echeverría led to increasing state intervention in rural production (Fox 1992), including timber production. Echeverría displayed the political will to expropriate forest lands from both small regional logging companies, such as in Guerrero, and from one of the few politically powerful Mexican timber companies, Forests of Chihuahua (Bosques de Chihuahua), shifting control to parastatals (see chapter 2). His reforms continued to seriously limit the role of international and national capital in timber and introduced a new policy current of state subsidies for the sector. Fox, in an analysis of food policy from 1980 to 1983, argues that reforms within the state emerge from "policy currents" and alliances between the state and social actors whose political efforts result in reforms (Fox 1992; Silva 1994). In the case of forests, the limited evidence that exists suggests frequent struggles between state agencies over the direction of policy in forestry. For example, the Secretary of Agrarian Reform commonly advocated for forest communities and provided support for the organization of CFEs in conflict with other federal agencies, state governments, and local economic interests. However, executive power, in the person of the president of Mexico, repeatedly intervened in these disputes on the side of forest communities (García-López and Antinori 2018; interview with Roberto Vidaña, February 12, 2018, Durango). The reforms of the early 1970s were clearly driven by fears that guerrilla uprisings in Chihuahua and Guerrero could spread. However, paired with military suppression were major forest land distributions to local communities as a rural pacification strategy. The role of "intermediation" by civil society in agrarian and forest policy in Mexico has been emphasized recently by P. Kashwan (2017), but more relevant are consistent state reforms and programs from above (see chapter 2). The reforms in the forest sector opened up not only a political space where rural people could demand more accountability from the state (Fox 2007; Kashwan 2017), which it did, but an institutional and economic space insulated from national and foreign capital and with property rights over a valuable natural resource, a "safe operating space" (Anderies, Mathias, and Janssen 2019; Dearing et al. 2014). J. Fox notes that the food reforms of the early 1980s did not lead to lasting change. However, the truly amazing thing is that forest sector reforms *did* lead to lasting structural change in an entire sector of the Mexican economy with social, economic, and ecological multiplier effects. These lasting effects contrast with the common characterization of the history of underdeveloped rural areas in the capitalist world-system as one of "ecological distribution conflicts" in which local communities are forced into "resistance" because the accumulation of capital degrades nature and communities on the periphery of the modern world economy (Hornborg 2007; Martínez-Alier 2004; Orta-Martínez and Finer 2010). García-López (2018) also argues that in Mexico "persistent democratic deficits reproduced by techno-bureaucratic forestry and authoritarian-corporatists logics" have negatively impacted CFEs, but this characterization misses the most salient aspects of Mexican CFEs. The Mexican institutional framework and initiatives from above and below significantly resolved ecological distribution conflicts in favor of forest communities with few democratic deficits. For decades communities have spent much more time running a business with state policy support than in resistance to capital accumulation or authoritarian-corporatist logics. Secure forest tenure gave communities natural capital, and varying state investments in the other four capitals created both a defensive perimeter against the deleterious impacts of global capitalism and a viable channel to engage with national markets in Mexico without being exploited. In this sense, Mexican community forestry is "counterhegemonic," as characterized by Alatorre Frenk (2000), and are unusual firms in a capitalist economy (see chapter 6). With this background, I will present the conceptual framework used in this book for the institutional framework of the resilient SES of Mexican CFEs, rooted in public policy, markets, communities, and political economy.

THE SOCIAL-ECOLOGICAL SYSTEM OF MEXICAN COMMUNITY FORESTRY: INSTITUTIONS FOR COLLECTIVE ACTION FROM ABOVE AND BELOW

Most of the work on SES is on local-level units such as lakes, forests, and small-scale fisheries (Leslie et al. 2015; Liu et al. 2007). However, with Mexican community forestry, we are dealing with a national spatial scale and a temporal scale of some one hundred years, forcing adaptations to the framework. The integrated SES framework proposed here is drawn on the preceding theoretical discussion, provides the structure for the chapters that follow, and allows for a holistic understanding of the remarkable historical achievement of Mexican community forestry. The emergence of the concept of SES arose as an effort to integrate the social, natural, and physical sciences into a single analytical framework. This effort to break down disciplinary boundaries addresses complex and "wicked" environmental problems that show no respect for academic disciplines (Bruggemann et al. 2012; Olsson et al. 2004). It has been proposed that SESs are deeply interconnected and coevolve across spatial and temporal scales, or cross-scales, and as complex adaptive systems, display nonlinear feedbacks, varying temporal processes, and the evolution of cooperation and social norms in a process of self-organization (Collins et al. 2011; Folke 2006; Levin et al. 2013). With some recent exceptions, discussions of complex adaptive systems commonly lack any sense of agency (Hahn and Nykvist 2017) and will not be further pursued here, since the Mexican forestry SES has clearly defined agency through rules and actors. As well, discussions of SES governance tend to have a strong normative component: "the emergence of such governance systems is often facilitated through enabling legislation, economic incentives, and by bridging organizations that connect institutions across levels and scales to enhance their capacity to deal with change" (Folke 2006, 15). It is not clear just how "often" such desirable outcomes are actually facilitated in the real world, and it is evident that any given SES is as likely, or more likely, to produce negative social and ecological outcomes than positive ones. Nonetheless, it is proposed in this book that across hundreds of cases in Mexico a resilient multilevel governance system has indeed emerged with substantial capacity to deal with change.

Elinor Ostrom was a pioneer in thinking about linked social and ecological systems, in her separation and examination of the interactions between a "governance system," the institutional arrangements, and the biophysical CPR, though the design principles themselves make little actual reference to ecology. However, in her later work within the Institutional Analysis and Development (IAD) framework, she broadened the focus to include ecological variables such as tree growth, forest cover, forest protection, and biodiversity conservation (Ostrom 2005, 2009; Ostrom and Nagendra 2006). Following on this exploration, in a series of publications, she created a formal SES framework (Ostrom 2007, 2009; McGinnis and Ostrom 2014) in which nearly all the variables are local and the framework is designed to "identify relevant variables for studying a single focal SES" (Ostrom 2009, 420). That framework thus has relatively little utility in studying a large-scale national SES such as Mexican community forestry and is arguably too tight a focus to account for the dynamics of most real-world SESs today. However, Ostrom contributed to a study of globalization of SES (Young et al. 2006), and the concept of "telecoupling" (Boillat et al. 2018; Liu et al. 2013) was developed precisely to insert the study of SES into larger social, economic, and political variables as key drivers of local systems. As well, there has recently been work on governing large-scale social-ecological systems (Cox 2014; Fleischman et al. 2014). F. D. Fleischman et al. (2014a) examined the applicability of Ostrom's (original unmodified) design principles across five large-scale SESs: forest management in Indonesia, the Great Barrier Reef in Australia, the Rhine River in Western Europe, the Ozone layer (i.e., the Montreal Protocol), and Atlantic bluefin tuna. Ostrom did not think the design principles were appropriate to use for more recently created and large-scale common property systems (H. Nagendra, personal communication, March 29, 2018) but, nonetheless, the comparison of these large-scale systems with Mexico is useful. The researchers find varied support for the design principles at larger scales. For example, fishers in the Great Barrier reef do not create their own harvest rules, and yet governance is successful; blue fin tuna fishers don't devise their own institutions, but can lobby and that "bottom-up self-organization may be difficult or impossible to achieve in large-scale systems" (Fleischman et al. 2014). The relevance of six other variables considered important in the small-scale SES literature are also proposed: dependence on the resources, group size / heterogeneity, external disturbances, resource characteristics, political power and civil society, and scientific knowledge. However, this evaluation does *not* take into account the four variables that I have proposed as central to the evolution of Mexican community forestry—state policy, markets and price incentives, the role of communities in bottom-up self-organization of CFEs, and political-economic reforms—which together have created resilience in the sector. Although Fleischman et al. (2014) note the salience of "political power," the discussion focuses on the role of civil society, when in all the cases discussed it appears that there is an overwhelming impact of state policy on management. As well, there is little appreciation for the role markets play in large-scale SES, and Mexican CFEs strongly contradict the finding that bottom-up self-organization may be impossible to achieve in large-scale systems (Fleischman et al. 2014, 451). It is, however, highly relevant to the Mexican case that Great Barrier Reef fishers and Atlantic bluefin tuna fishers do not create their own institutions and yet governance is successful. I would propose that this similarity lies in the high market value of the catch and the high market value of timber.

A final introductory comment: it is urgent that climate change be incorporated into conceptual thinking on SESs. The concept of SES as such is rapidly becoming antiquated while climate change progressively impacts both social and ecological systems. We must begin thinking in terms of social-ecological-climate systems (SECS), an exploration of which has barely begun (Fedele et al. 2019). Figure 1 is presented as an SECS, where the entire system results in both resilience, at the bottom of the figure, and adaptation and mitigation to climate change, at the top of the figure. Nonetheless, for the sake of simplicity and because climate change is not yet having a major impact on Mexican community forestry (but it is coming—see chapter 8), I will confine this book's discussion to using the concept of SES.

With that background, figure 1 outlines a conceptual framework for the SES of Mexican community forestry and the organizing framework for the book. As proposed by Ostrom (2011), this book's framework provides the most general set of variables, which I propose allow us to understand the dynamics of Mexican

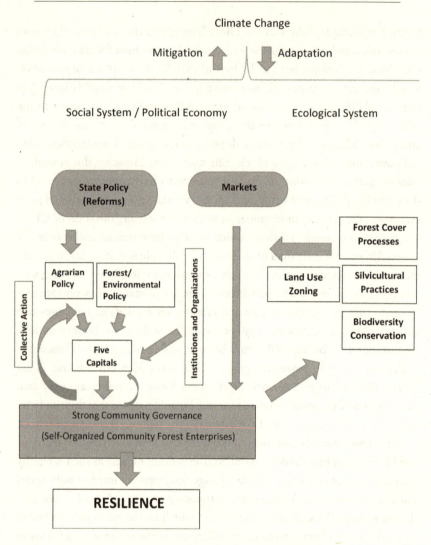

FIGURE 1. The social-ecological-climate system of Mexican CFEs.

community forestry, and it also represents a theory. State policy, in both supporting programs and reforms, and markets in interaction with increasingly autonomous community collective action have created the unique properties of the sector.

In Mexico, the institutional framework for collective action by the CFEs was initially created, in successive historical events over most of the twentieth and into the twenty-first century, by state policy reforms and in later periods by CFE engagement with markets. Hence, state reforms and CFE-market interaction are the principal drivers of the system. As we have seen, the village and territorial governance platform was provided by agrarian policy and the emergence of CFEs later was provided by forest and rural development programs and policy. The endowment of the five capitals, from both forest and rural development policies, created the institutional framework that provided incentives for strong community collective action (Bray, Merino-Pérez, and Barry 2005b), first for forestland redistribution and later for support for CFEs, which in turn influenced state policy. Thus, there are feedback loops from strong community governance and social capital driven by profits from the CFEs that encouraged further policy reforms and strong links to markets. As Levin et al. (2013, 116) have noted for SESs, "Policy creates social frames that act as attractors and help focus the market's emergent properties towards a desired goal." In the case of Mexico, the social frame of agrarian, development, and forest policies helped channel an evolution toward communities gaining access to their tree rights and to diversifying beyond being forced to sell to a single buyer, ending up in a more competitive buyer's market that drove deepening collective action around CFEs. This process shows that although community governance was historically heavily influenced by exogenous variables of government policy and markets, it is now a central and more autonomous endogenous variable driving the system.

The arrow on the left from community governance to policy, labeled collective action, represents both historical periods of community and intercommunity mobilization against periodic state abuses and, in more recent decades, general policy support for CFEs, given its demonstrated success. The interaction of top-down and bottom-up forces constituted the institutions and organizations of the sector (vertical bar on right). This framework presumes the best-case scenarios of successful commons management, of which many hundreds of cases exist in Mexico. Nonetheless, the incentives toward a communal enterprise have not always been strong enough to overcome disincentives, and some CFEs have broken down into subgroups or subcoalitions or individual production and, in likely hundreds of other cases, corruption, elite capture, and poor forest management are present. In these cases, clearly the transaction costs of collective action around the forest were judged to be too high (see chapter 5).

Nonetheless, in its most positive manifestations, this social system / political economy has in turn had a strong impact in shaping the ecological nature of the community forests. The historical institutional evolution has brought ecological zoning to millions of hectares of Mexican forests. As we shall see in chapter 7, CFE communities are zoned into both forest production and conservation areas (as well as agricultural and village areas). This zoning represents the social system driving the ordering of the ecological system, and it is shown in figure 1 as the essential link between the social and ecological systems. Territorial governance practices usually classify significant forest areas as under conservation, which impacts biodiversity conservation. In the forest production areas, the structure and composition of the forest are strongly influenced by the silvicultural practices used. As a result, the overall structure, composition, and forest cover, at the level of both the community territory and entire regions, is impacted by these territorial silvicultural governance practices, since CFEs are normally clustered together in areas with good commercial forests. Positive results in terms of timber production, reduction of deforestation, and biodiversity conservation have also created a positive feedback loop to encourage more government programs that support these activities, represented in figure 1 by the arrow pointing from the ecological system to the social system. It is the central thesis of this book that this coupled social-ecological system of forest commons management has been made more resilient by external "forcings" (to borrow a term from climate science) such as markets and government policy and that it has developed, in many cases, a "capacity for permanent readaptation" (Alatorre Frenk 2000, 17) or adaptability. This capacity for permanent readaptation has likely made the forest ecological system highly resilient to climate change, a subject to which I shall return in the concluding chapter.

The conceptual framework is exploratory and suggests potential relationships among variables. Causal relationships would depend on more detailed empirical studies. This framework also provides the structure of the remainder of the book. In chapter 2, expanding Ostrom's notion of facilitative regimes and institutional supply from this chapter, I will show how the evolution of government agrarian, development, and forest policies from the Mexican Revolution through 1988 laid the foundation for the development of CFEs with endowments of the five capitals and with periodic social mobilizations in particular regions demanding expansion of the spaces. Chapter 3 discusses how the further evolution of these policies from 1988 to 2018 constituted a period in which CFEs normally no longer had to struggle for space to exist, but set about learning how to operate their businesses, as well as provides a review of selected aspects of the community forest sector today. In chapter 4 we see how community governance of the CFEs has evolved from varying bases of disorganization and organization, with two comprehensive case studies of CFEs with contrasting origins. In chapter 5, I review the institutional dimensions of CFEs and how they have evolved. Chapter 6 examines issues of competitiveness and profitability in CFEs. Chapter 7 examines the ecology of the SES and how CFEs are controlling deforestation and conserving biodiversity through territorial zoning practices. Concluding chapter 8 reviews the state of this SES today and how it is adapting to the new stressor of climate change from higher precipitation and temperatures and extreme storm events, increasingly driving threats to forest livelihoods and loss of biodiversity (Estes et al. 2011; Rands et al. 2010). I will argue that Mexican community forestry constitutes an example of what has been called the seeds of a good Anthropocene, a sustainable pathway for increasing the resilience of the forests of the developing world (E. C. Ellis 2018). The Mexican experience can also inform broader policy contexts related to the proposed global program for REDD+ (Reduction of Emissions from Deforestation and Degradation "plus" conservation, sustainable forest management, and enhancement of forest carbon stocks) (Angelsen et al. 2009). Many community forest regions of Mexico represent what could be called "post-REDD+" landscapes, what other parts of the world might look like if REDD+ were successfully implemented (Bray 2010a). State policy, markets, and communities as sites of entrepreneurial collective action have converged to create the large-scale, temporally dynamic, spatially dispersed, and resilient SES of Mexican community forestry, which continues to be vibrant at the end of the second decade of the twenty-first century, despite serious problems in some regions and states.

CHAPTER 2

THE ORIGINS OF MEXICAN COMMUNITY FORESTRY

Facilitative Political Regimes and Community Collective Action

I n this chapter, I will cover the ecological foundations of the Mexican community forestry SES, the history of institutional supply, the dynamics of a facilitative political regime, and community collective action up to the 1986 Forest Law and the end of the period of President Miguel de la Madrid in 1988. I briefly introduced the concept of institutional supply in chapter 1, but it requires some further explanation for this chapter. Ostrom divided the problem of institutional supply into the origins of rules and the analysis of changes in existing rules. Her classical cases of CPR management in the mountain forests and pastures of Switzerland and Japan and irrigation systems of Spain and the Philippines characterized the origins of the institutions as "lost in time" (Ostrom 1990, 103), and she focused on existing rules at the time of the study. For the origins of governance institutions, as noted in chapter 1, she turned to groundwater basins in the 1950s Los Angeles metropolitan area to analyze the litigious process of creating rules around groundwater access in a rapidly growing region. She suggested that "the origin of institutions is thought of as a major one-step transformation, whereas institutional change is viewed as involving incremental changes in existing rules. Supplying new institutions is consequently viewed as non-incremental and costly, whereas changing institutions is viewed as incremental and not as costly" (Ostrom 1990, 140). The Mexican agrarian revolution (1911–17) first created the supply of new institutions for community governance and became the foundation for community forest management. However, as we will see in this chapter, it was a much more challenging, costly, and violently nonincremental event than a couple of decades of litigation over groundwater rights in Los Angeles.

Before turning to the history of institutional and capital supply in Mexico, I will briefly describe the ecological dimension of the SES that has been molded by decades of institutional supply. Mexico's enormous diversity in vegetation and related biodiversity has resulted in classifications with up to fifty vegetation types. Nonetheless, there are eight that have been proposed for general use: evergreen tropical forests, semideciduous tropical forests, deciduous or dry tropical forests, *bosques mesófilos de montaña,* temperate coniferous and broadleaf forests, dry *matorrales*, pastures, and wetlands (Challenger and Soberón 2008). I propose modifications to two of these terms. The concept of *bosques mesófilos de montaña* comes from Mexican botanist Jerzy Rzedowski (1988) but is not recognized outside of Mexico, and it is interpreted to include both montane tropical forests (between 1,000 and 2,500 meters) and tropical montane cloud forests (TMCF), above 2,500 meters (L. S. Hamilton et al. 2012). For temperate coniferous and broadleaf forests, I will refer to pine-oak forests or pine forests. Map 1 shows the historic distribution of the major forest types. (See chapter 7 for map 2, which shows the distribution as of the 2010s.)

There is little management of dry tropical forests, and the arid and semiarid *matorral* vegetation, pastures, and wetlands are not relevant components of the community forest SES, so in this book I will discuss only evergreen and semideciduous tropical and temperate forests. There is also virtually no community forest management for timber in tropical montane and TCMF forests, though they will be relevant in chapter 7 in regard to the conservation and ecology of community forests. TMCF forests originally occupied only 3.09 million hectares but today have been reduced to around 870,000 hectares in primary condition and 950,000 in secondary, for a mere 0.83 percent of the national territory. Despite this reduced area, it has 9 percent of national floral diversity with about 650 genuses and the highest concentration of plant biodiversity by surface unit in Mexico (Challenger and Soberón 2008).

Evergreen or lowland tropical forests (below 1,000 meters) are found principally along the Gulf Coast plains, in the southern and eastern Yucatán Peninsula, and in eastern Chiapas (the Lacandón rainforest). On the Pacific Coast, tropical forests are along the coast of Chiapas, and in lower elevations in the Sierra Madres of Oaxaca and Guerrero. These forests have annual rainfall of

Potential or Historic Vegetation-Mexico

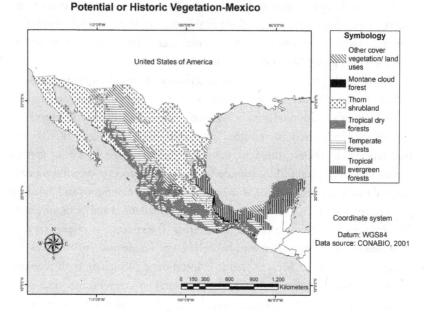

MAP 1. Historic vegetation cover—Mexico. Translated from J. Rzedowski (1990).

2,000 millimeters or higher and originally covered 9.1 percent of the national territory. Today they have been reduced to a little more than half that, 4.82 percent (distributed between 3.16 million hectares of primary forest and 6.31 million hectares in secondary forest), and have 17 percent of the plant biodiversity of Mexico, at around 30,000 species (Challenger and Soberón 2008). Tropical forest management is confined almost entirely to the semideciduous forests of central Quintana Roo (with around 1,000 meters of annual rainfall) and the wetter forests of southern Quintana Roo and southern Campeche in the Yucatán Peninsula.

The basic armature for Mexico's pine-oak forests and montane tropical forests is in the Sierra Madre Occidental and the Sierra Madre Oriental, the mountain ranges in western and eastern Mexico, the Volcanic Axis that joins the two ranges in central Mexico, and the Sierra Madre del Sur along the Pacific Coast of Guerrero and Oaxaca. In the south, after a break at the Isthmus of Tehuantepec, the mountains rise again in the Sierra Madre de Chiapas and the Mesa de Chiapas in southeastern Mexico. The pine-oak forests historically occupied 43.96 million hectares and today occupy 16.45 percent of national territory (21.19 million hectares in primary forest and 11.13 million hectares in secondary). Pure stands of oak, relatively little used commercially, are usually found at the mid-elevations, with mixtures of pine and oak higher up and pine occurring in pure stands at higher altitudes. Mexico is the major center of diversity for both pine and oaks globally, home to 46 of the 110 pine species and 170 of the 450 oak species known, and it is considered the major center for evolution and diversity of these species in the Americas (Galicia, Potvin, and Messier 2015; Challenger and Soberón 2008). Most community forest management experiences occur in the Sierra pine-oak forests with Chihuahua, Durango, Michoacán, Guerrero, Mexico State, Puebla, and Oaxaca the most important states and with smaller numbers in the tropical forests of Quintana Roo and Campeche, as noted. Temperate and tropical forests provide habitat for much of Mexico's vaunted biodiversity. Of the seventeen most "megadiverse" countries globally, Mexico places fourth, much of it in the forests. Mexico is second in the world in diversity of reptiles and is among the top five in amphibians, mammals, and flowering plants (Espinosa and Ocegueda et al. 2008). The temperate forests have recorded 656 species of amphibians, 1,300 bird species, 1,586 reptile species, and 146 mammal species (Challenger 1998). I will expand on the ecology of Mexican forests in chapter 7.

THE ORIGINS OF INSTITUTIONAL SUPPLY FOR COMMON PROPERTY GOVERNANCE OF MEXICO'S TEMPERATE AND TROPICAL FORESTS

The common property system that emerged after the Mexican Revolution was an indirect descendant of the Aztec *calpulli*, a kinship-based communal land tenure system (Wilson and Thompson 1993). Indigenous communal institutions then went through three centuries of evolving Spanish colonial redesign of property rights, including the important legal principal that common lands could not be alienated (Simpson 1937). In practice, however, over half of the putatively legally recognized indigenous common properties in the colonial period were absorbed into haciendas by the end of the seventeenth century, aided by the earlier disease-driven demographic collapse of the native populations. But some common property institutions survived, particularly in Oaxaca, where mountain isolation allowed communal governance traditions to persist

(Assies 2008), though now channeled through the colonial *municipio* institutions (Bailón Corres 1999). The term *ejido*, imported from Spain and codified in a 1573 law, referred to community pastures (W. B. Taylor 1972), and it was not until the nineteenth century that this term evolved to refer to all community common lands (Assies 2008; Knowlton 1998).

Mexico gained political independence in 1821, and the ensuing chaos of the first decades basically froze most property rights situations. A liberalizing impulse produced the Ley Lerdo of 1856, principally aimed at ending widespread church ownership of land, but also defining indigenous commons as an impediment to progress. The law ended the rule of inalienability of communal lands, with the goal of achieving a class of free small farmers (Assies 2008). This legislation would not have its full impact until the period of the dictator Porfirio Díaz (1876–80, 1884–1911), known as the Porfiriato, who carried out a vast redistribution of communal and government lands to mostly foreign owners. Many communities with customary or legal land rights lost possession through purchase, destruction of titles, or expropriation. For example, a Tarascan village in Michoacán with a colonial-era title over an extensive marsh area lost its rights by dubious means to Spanish investors (Friedrich 1970). Processes such as these, multiplied thousands of times, meant that by the end of the Díaz period in 1911 some 87 percent of the land was held by 0.2 percent of the landowners and 95 percent of rural heads of households were landless (Assies 2008; Markiewicz 1993). The exception again was southern Mexico, which remained insulated from the worst of these depredations.

Forests during the Porfiriato were also handed out to foreign interests, and U.S. investors entered Chihuahua, Durango, and Michoacán in the late nineteenth and early twentieth centuries, receiving concessions over vast forest estates that included many traditional community lands (Boyer 2015). In the southern tropical forests, a centuries-old trade in mahogany and Spanish cedar gained new impetus in huge new concessions (de Vos 1996). Deforestation emanating from Mexico City and other urban areas in the late nineteenth and early twentieth centuries spurred Miguel Angel de Quevedo, a French-trained hydrologist, to begin to agitate for more forest regulation. De Quevedo established the first forestry school in Mexico and the first federal agency dedicated to forests in 1908. In 1909 he attended a forest conference in Washington, D.C., invited by the first chief of the U.S. Forest Service, Gifford Pinchot. These incipient efforts at state forest policy were swept away by the Mexican Revolution (Boyer 2015; Simonian 1995).

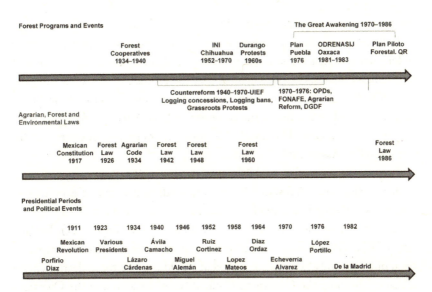

FIGURE 2. Mexican forest policy history timeline (1917–86). Adapted from Mathews (2006).

For reference, figure 2 outlines the history of Mexican forest policy that will be discussed in the remainder of this chapter, showing the major legislative and other events that defined the period, terminating with the 1986 Forest Law and end of the period of President Miguel de la Madrid in 1988.

The Mexican Revolution (1911–17) began as a middle-class protest for fair elections but was quickly eclipsed by rural grievances over the massive loss of property and communal rights in the Porfiriato. The Plan de Ayala, promulgated by peasant revolutionary leader Emiliano Zapata in 1911, called for the immediate return of community lands, with few immediate results. Elite landowners regained political control by 1917, but in response to Zapata's demands and the rural uprisings, Article 27 of the Constitution of 1917 established that all lands, forests, and waters originally belonged to the nation, to be distributed as both private and communal property, with de Quevedo succeeding in getting a reference to conservation inserted (Boyer 2015; Sanderson 1981; Silva Herzog 1959).

However, the architects of the agrarian reform saw the reconstitution of common property as a stopgap measure to end the rebellion and as a way station to small private property (Sanderson 1981). The next two decades would see ongoing struggles over the constitutional and collective choice rules for property in Mexico, so the emergence of the contemporary common property

regime was not a historical inevitability. The 1917 Constitution established a lit-
eral constitutional choice for government-regulated communal properties given
in long-term usufruct. The actual distribution of land was exceedingly slow as
large landed interests asserted control over the process. In the 1920s and early
1930s, the debate raged over whether communal rights were a transitional form
to private property, favored by the conservative landowners who controlled the
government, or a new and uniquely Mexican form of property, neither socialist
nor capitalist (Assies 2008; Fox 1992).

Legislation in 1922 first established the two forms of agrarian property mentioned in chapter 1: ejidos, distributions to landless peasants of expropriated
haciendas and state lands, and comunidades, the restoration of lands to indigenous communities with titles from the colonial period (Boyer 2015). The 1925
Law of Ejido Patrimony is crucial as the first expression of the contemporary
form of community governance and what would become the collective choice
arena, establishing the Governing Council (Comisariado) and the Oversight
Council (Consejo de Vigilancia), with the requirement of democratic election
by majority vote in the Assembly of all legal community members. The elected
officials represent the village to other levels of government and manage the
common property, among other duties (Simpson 1937). The legislation built
a "nested enterprise" of multilevel governance by establishing an Agrarian
Department with the powers to closely regulate the ejidos and comunidades.
The property rights granted to agrarian communities had the primary purpose
of preventing peasant revolt and for this reason were severely restrained. Land
rentals and sales were prohibited, dividing the commons was illegal, proposals
for new members had to be approved by the Secretary of Agrarian Reform, and
inheritance rights were granted to only one descendant, overall a paternalistic
subordination to state controls (Ibarra Mendivel 1996).

In the meantime, de Quevedo attempted with colleagues to establish an
apparatus of forest regulation with a 1926 Forest Law. This law was a reaction
to the emerging distribution of forests as well as agricultural land, with de Quevedo opposing the distribution of forests to "ill-prepared" communities, though
the law quickly became a "dead letter" (Boyer 2015, 98). The law did, however, call
for the establishment of community forest cooperatives and the requirement of
logging permits based on management plans, features that became relevant in
the 1930s. At the beginning of the period of Mexico's reformist president Lázaro
Cárdenas (1934–40), the global Depression and the return of migrant workers from the United States forced an expansion of land distribution, including

forests. During this period, institutions of communal land tenure and governance became a permanent and uniquely Mexican form of common property, a contribution of social capital, and the distribution of natural capital was launched, the first two of the supply of five capitals that drove the emergence of CFEs.

In 1933, rules and organization for agrarian administration were defined in Article 27 of the Mexican Constitution, codifying them as an imposed constitutional choice. The 1934 Agrarian Code affirmed a top-down collective choice rule, stating that the Comisariado is composed of a president, a secretary, and a treasurer democratically elected by the Assembly along with the Oversight Council (Simpson 1937). It established a land-zoning model, a crucial element in forest management, which defined village residential areas, individually worked agricultural areas, and common pastures and/or forests, and it stated that communal lands were "imprescriptible and inalienable (and cannot be) ceded, conveyed or mortgaged or made subject of lien in whole or in part" (324). Students of Mexican agrarian reform have emphasized the process of land redistribution, but it is also crucial to emphasize the signal importance of the establishment of clear rules for the institutional and organizational forms of governance. This national top-down framework was certainly conceived as mechanism of control by the state. However, it also provided the state-structured space for self-governance at the community level. It was the first major step of the transformative decentralization that would result in an economic defensive perimeter and long-term processes of "adaptability" that led to CFEs. Two levels of institutional supply were provided, clearly linking communities to the state and providing a national template of rules for community governance that specified rights of access, withdrawal, management, exclusion, and alienation, reducing transaction costs for the community, who did not have to create all these rules themselves. But it was a long historical process via which rural Mexican citizens have been able to use this imposed framework to provide Ostrom's basic level of institutional supply, the operational rules such as monitoring and enforcement. Further, as we shall see, forest harvest rules—appropriation and provision in Ostrom's terms—would also be supplied from above, in the environmental and forest laws.

Supplying the governance institutions was clearly long, bloody, nonincremental, and costly. By 1934 the law had been applied in fewer than 7 percent of the then-existing 4,069 ejidos, leaving most communities with "every variety and degree of organization and disorganization" and left to work out "their own individual systems" (Simpson 1937, 342–43). Most agrarian communities were ruled by authoritarian caciques and were still sunk in the violent chaos of the revolutionary period, a status quo of "rulelessness." For example, for the state of Hidalgo "the organization and administration of the ejidos will require a long time; the work is obstructed by . . . the terrible heritage which these dissensions, many of them sealed in blood and only to be resolved by death itself, have left among the *ejidatarios*" (346).[2]

Joined with the agrarian reform policies, the forest policies of President Cárdenas were modeled on his prior service as governor of Michoacán, representing the first clear departure from the prerevolutionary practice of handing over vast forest areas to foreign investors. In 1931, while still governor, Cárdenas had observed the rapid deforestation in Michoacán, declared null and void all timber contracts in the state, and required community cooperatives for logging (Boyer 2015; Hinojosa Ortiz 1958), the first policy step towards CFEs in Mexico. As president, Cárdenas established the cabinet-level Autonomous Department of Forests, Game and Fisheries led by de Quevedo. Cárdenas took full advantage of the call for forest cooperatives in the 1926 law, and by 1940 forest extension agents had helped organize 866 forest cooperatives with management plans, 64 percent of the estimated 1,350 forest communities at the time. As Boyer (2015, 123) has noted, "the policies that Cárdenas and de Quevedo put in place constituted the first serious effort to find a more sustainable and equitable use of the nation's forests . . . [T]he forest service . . . had set as its goal the management of forests on a national scale based in part on the ideal of local management." However, the cooperatives were primarily a fiction to conceal logging by contractors with little community participation (Boyer 2015; Hinojosa Ortiz 1958). Nonetheless, it is also in this period that some early enormous grants of natural capital were handed out and that at least some of the cooperatives were also endowed with sawmills, their first physical capital, even if operated by the contractors with community labor. Leaders of the community of Pueblo Nuevo in Durango demanded the restitution of land taken in the nineteenth century, and it received its first enormous land grant, of 166,754 hectares, in 1931 (Guerra Lizárraga 1991). A cooperative was established later in the 1930s, and a private sawmill was installed where community members began to learn the basics of sawmill operations (Hernández Astorga 2015). Today it is one of Mexico's largest CFEs, with a total of 243,349 hectares. Similar processes occurred in the

2. An *ejidatario* is the Spanish term for an ejido rightsholder.

case of the Chihuahua communities of Samachique and Guachochi and communities in Quintana Roo, which evolved to manage their own autonomous CFEs in recent decades (Boyer 2015; Ureña Argaez 2017). The beginnings of the distribution of forest lands and the cooperatives, as flawed as they were, laid the foundations for many successful CFEs to emerge decades later. As well, in this period, grassroots movements to demand land titles for forest lands emerged, taking advantage of the promise of agrarian reform. In addition to Pueblo Nuevo's grant, other early enormous endowments of natural capital in Durango in the 1930s founded more than twenty other forest ejidos including such now well-established CFEs as San Pablo, El Brillante, and La Ciudad (Guerra Lizárraga 1991; Hernández Astorga 2015).

Despite the forest grants, the Cárdenas-era cooperatives represented little advance in developing new governance institutions for the establishment of CFEs or training to form human capital for the entirely novel and nontraditional practice of industrial forestry, though some community workers received basic training in industrial forestry from the contractors. In nine states, Cárdenas also launched an enduring policy of instituting logging bans to halt deforestation (Boyer 2015). In addition to these early efforts at promoting both CFEs and bans, Mexico also became a world pioneer in the establishment of protected areas. The Cárdenas administration established forty national parks, more parks than existed in any other country in the world at the time, comprising two million acres of land in fourteen states. These parks also pioneered early concepts of "multiple-use," since farmers continued to use them to grow crops and produce charcoal while recreational uses emerged (Wakild 2011). Thus, in Mexico at the time, community forests and protected areas were not seen as opposing alternatives, but as complementary.

When the conservative government of President Manuel Ávila Camacho came to power in 1940, the period of forest cooperatives, as inadequate as it had been, was ended formally by suppressing the law in 1943. The 1940–70 period, encompassing five Mexican presidencies, is known in Mexican political history as the "counter-reform" (Sanderson 1981), and community forestry would be almost entirely ignored in the name of economic development, with the notable exception of efforts on the part of the national government in Chihuahua in the 1950s and community struggles to form the first real CFEs in Durango in the 1960s, to be discussed below. However, at least in the early 1940s, major extensions of natural capital continued to be endowed, for example, in the Durango communities of La Ciudad in 1940 and Vencedores in 1941, though new land grants were given out slowly under pressure from the Durango Lumber Company. As well, the first documented effort to establish an independent CFE in Pueblo Nuevo ended with the assassination of its leader in March 1944 (Guerra Lizárraga 1991). This vision for a CFE was not to be regained in Durango until the mid-1960s.

During the counter-reform, the Mexican state would promote three coterminous and frequently contradictory policy initiatives, while a fourth underlying trend began to impact policy. The three policies were (a) harnessing Mexican forests to import-substitution industrialization for paper linked to a policy of granting vast logging concessions to parastatals and private companies; (b) implementing bans as an effort to halt clandestine logging, particularly in areas without large economic interests; and (c) in the 1950s, in an expression of the persisting ideals of the revolution, isolated efforts in Chihuahua and Chiapas to train local communities to manage their own CFEs. Finally, a fourth significant trend in this period was the emergence of grassroots protests and resistance to logging on their lands by communities and calls for the creation of CFEs, particularly in Durango in the 1960s.

IMPORT-SUBSTITUTION INDUSTRIALIZATION AND CONCESSIONS

The first major state-centralized efforts at logging were the Industrial Units for Forest Exploitation (Unidades Industriales de Explotación Forestal; UIEF), stimulated by paper scarcities during World War II. Industrialists acted to assure their supply of timber and to roll back the establishment of cooperatives and national parks promoted under Cárdenas (Aguilar Espinoza 1990). A 1943 Forest Law established the creation of the UIEFs, though rules were not codified until the late 1940s. The UIEFs gave the government authority to turn over the rights to the trees on community lands and on the still-extensive state lands to industry. About half of Mexico's forests were placed under logging concessions, thirty-seven to parastatals, and ten to private companies (García-López 2012), stripping the communities of tree rights in a "partial expropriation" based on Article 27 of the Mexican Constitution (Griffith 1958, 31), with the communities to be paid an administratively set stumpage fee (*derecho de monte*) (Chambille 1983). The companies also had to pay for a bureaucratic apparatus of technical supervision and management. Eventually twenty-one UIEFs were established in the 1940s and 1950s, though nearly half of them were cancelled due to mismanagement and resistance from communities (Boyer 2015).

With the demise of the Cárdenas-era cooperatives, the principal mechanism for linking contractors and communities became the buy-sell contracts (*contratos de compra-venta*). According to these contracts, all logging costs were assumed by the buyer and preference in employment was to be given to the community, though commonly the companies brought their own lumberjacks. The stumpage fee was paid directly to a fund administered by the Secretary of Agriculture, and it was frequently difficult for the communities to access. From 1942 to 1946 there was an annual average of 450 compra-venta contracts approved in ejidos and comunidades (López Santos 1948). In this period, production from community lands was relatively unimportant, constituting less than 20 percent of national production, with over three-fourths coming from state-owned and private lands (Hinojosa Ortiz 1958).

BANS

There was widespread concern about advancing deforestation in many regions of Mexico, and logging bans were the favored policy to deal with it. Between 1940 and 1952, bans were declared in twenty states, covering vast extensions of territory. In Durango, two million hectares were placed under a ban, and in Chihuahua for a period in the late 1940s, the entire state was under a ban (Zarzosa 1958). Overall, some 60 percent of Mexican forests were under bans and other forms of protection in the 1950s (Boyer 2015). According to one observer, the bans were less about conservation than a means to punish political enemies (interview with former high-ranking forest official, December 6, 2016). However, they seldom had the desired effect. In the Cofre de Perote region of Veracruz in 1958,

Very few inhabitants of the mountain were informed of the existence of the ban. The forest exploitation didn't diminish; to the contrary . . . "Don Raul" (a local cacique) took charge of finishing off what remained of the woods. His work gangs worked under great pressure in all the regions, free of the vigilance of the forest guards. The latter, on the other hand, detained, fined and relieved of their timber and saws the peasants who went down to sell planks or beams in nearby cities . . . to avoid the forest guards they had to go down at night, whether it was foggy or rainy, by steep, muddy roads, to be able to return with food and money for family expenses. If the forest guards caught them, they had to give them a bribe, or be left with nothing and go back to their communities empty-handed. (Geréz Fernández 1993, 3)

THE RETURN OF STATE EFFORTS TO ORGANIZE CFES

Despite the counter-reform, the "recurrent reformist presence" in the state (Fox 1992, 40) persisted in expressing an alternative vision, that Mexican forest communities could one day manage their own forests. For example, a subsecretary of forestry in the 1950s commented:

> At the present time, it is impossible . . . to organize in Mexico cooperatives in the style of Switzerland, Finland or Sweden, but we can and should create organizations that, without turning their backs on the people, organize the exploitation for the benefit of comuneros and ejidatarios. These organizations should employ the labor of the members of the community and go educating and preparing them so that, in the future, they can approach and resolve all the technical, industrial, commercial and administrative problems that forest exploitation demands. (Hinojosa Ortiz 1958, 245)[3]

Beginning in the 1950s, for the first time since the 1930s, these recurrent top-down reformers made themselves felt, as a federal government agency began to build the community institutions and organizations for CFEs in Chihuahua, though now with a clearer focus on creating autonomy from the logging industry and on investing in natural, physical, human, and social capital than in the earlier decade.

In 1952, a center of the National Indigenous Institute (Instituto Nacional Indigenista; INI) was established in Guachochi, Chihuahua, and charged with intermediating between ejidos and comunidades and loggers in three highland municipalities (Boyer 2015). The Indigenous Coordinating Center of the Tarahumara (Centro Coordinador Indigenista de la Tarahumara; CCIT) began to support Tarahumara (now more correctly referred to as Rarámuri) communities to build capacity for management of their forests. However, a number of these communities were actually dominated by mestizos, a common problem in Chihuahua, so it is not clear how much the Rarámuri benefited (interview

---

3. All translations from the Spanish throughout the book are mine.

with former forest official, December 6, 2016). In any event, the CCIT first worked on land titling; securing natural capital, since many of the communities still did not have agrarian titles to their territories; suspending contracts with timber companies; and then supporting the communities in renegotiating those contracts. The first director of the CCIT, anthropologist Francisco Plancarte, reported on one session, "I have just told these Indians that they should not accept the lease agreement you propose. I explained that it amounts to extortion, and they agree with me. I invite you to improve your offer" (Boyer 2015, 177). The CCIT had the explicit goal of creating CFEs, where the communities would slowly begin to acquire the skills and governance capacities associated with industrial logging. This led to fierce bureaucratic and economic struggles between INI and the established government and private sector interests (interview with S. Nahmad, May 20, 1994). These struggles were the first in a pattern of conflict between federal agencies that supported community forestry and state and federal agencies that opposed it, allied with private industry. However, INI was able to carve out a political space to operate, and over the years it trained and equipped the communities to manage their own CFEs, and income from forests began to rise as a result (Boyer 2015). By the mid-1960s, after over a decade of sustained forest extension investments in human capital and endowments of natural and physical capital, various stages of incipient CFEs in sixteen communities in the Guachochi municipality had been achieved, and the Union of Ejidal Enterprises of the Tarahumara was organized (Boyer 2015). Cusárare, a surviving forest cooperative from the Cárdenas era, subject to elite capture in the 1940s, became the CCIT's showcase CFE (Boyer 2015), a clear example of the political economy of reform from above. The CCIT in Chihuahua represents the first concentrated effort by a Mexican government agency to create the institutions and organizations; secure natural capital; and invest in human, social, and physical capital that would allow for relatively autonomous administration of CFEs. The CCIT was thus the first significant, if regionally limited, step toward the transformation of the sector toward a new SES of well-governed community forests, opening spaces that were later occupied by vigorous community collective action.

GRASSROOTS PROTESTS

Chihuahua and, with less success, Chiapas, were the only states in the 1950s where significant top-down initiatives took place in this period. Elsewhere in

Mexico communities struggles from below were persistent from the 1920s on to achieve community land titles and greater autonomy in forest management. In the 1960s in Durango, Chihuahua, and elsewhere, state policy created an arrangement called "participatory associations" (*asociaciones en participación*, replacing the contratos de compra-venta of the 1940s and 1950s), which placed communities in a business partnership with logging companies but partnerships in which the seller was tied exclusively to the logging company. However, the reality of Mexican land tenure gave communities what economists call "hold-up rights" to resist logging in their territories, clawing back their tree rights, and communities increasingly refused to enter into contracts with logging concessionaires. Actions against logging abuses ranged from nonviolent, labor-oriented movements to guerrilla uprisings. In 1962 the communities of El Naranjal and San Vicente de Jesús in Guerrero forced small, private logging companies out of their forests. In 1964 three other Guerrero communities struggled against the UIEF Maderas Papanoa, a movement headed by a young schoolteacher, Lucio Cabañas, who led a significant regional guerrilla uprising in the early 1970s.

However, it was in Durango that the most forceful early and successful grassroots protests against concession logging and demands for forest lands occurred (interview with G. García-López, April 26, 2017). The Unión de Empresas Ejidales Forestales–El Salto (Ejido Enterprises Forest Union–El Salto) in the municipality of Pueblo Nuevo emerged in 1968 from a "liberation struggle" (García-López and Antinori 2018, 197) with strong grassroots leadership though quickly becoming supported by the government Confederación Nacional Campesina and the agrarian reform agency. This movement of recurrent reformers within the state apparatus encouraging and allying themselves with the grassroots protests was eventually supported by President Díaz Ordaz, a recurring example of executive power interceding in the support of the community forest sector.

In the 1960s the ejidos focused on demanding titles to territories that they occupied but that were under concession to the Durango Lumber Company (García-López and Antinori 2018). Victories in obtaining forest land grants in Durango and by INI in Chihuahua and elsewhere were part of the single largest wave of agrarian reform in Mexico, during the presidency of Gustavo Díaz Ordaz (1964–70). There are no statistics available on distribution of forest land specifically, but figure 3 shows the history of overall land distribution in twentieth-century Mexico, until new applications formally ended during the presidency of Salinas de Gortari in 1992. Land distributions before the Cárdenas

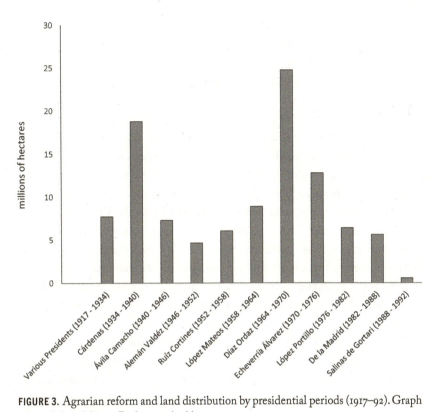

FIGURE 3. Agrarian reform and land distribution by presidential periods (1917–92). Graph adapted from Núñez Rodríguez (n.d.).

period were minor, but his presidency picked up the pace substantially, distributing some 18.8 million hectares, for nearly 10 percent of all distributions in the twentieth century. Distributions continued during the counterreform (1940–70) but at much reduced levels. However, President Díaz Ordaz ended up distributing 24.7 million hectares, for 12.6 percent of all distributions, so it is likely that 1940–70 was the most vigorous period for forest distributions as well. This finding is important because it shows that the natural capital base of the community forest sector is a result of historically recent reforms.

For example, from 1964 to 1970 some twenty-one new community land titles or expansions of existing communities were granted in the Pueblo Nuevo municipality after both vigorous grassroots mobilizations and pressure from the Alliance for Progress for land reform (García-López 2013; Guerra Lizarraga 1991). The struggles for land and organizational space were real, but it is highly noteworthy that the state usually ceded to the demands against local economic and political interests and later provided support for the further expansion of the community enterprise operating space. As noted, the 1960s wave of new land distributions in Durango and elsewhere were attributed at least in part to fears of rural rebellions stemming from the Cuban Revolution, the short-lived Ciudad Madera uprising of 1965 (see later in chapter), and pressures from the Alliance for Progress resulting from the Punta del Este Accords of 1961 "that conditioned financing credits from the USA on those countries . . . in Latin America that promote Agrarian Reform . . ." (Guerra Lizárraga 1991, 92). Land distribution in Durango in this period is the only reported case in which international pressures may have helped promote the CFE sector. Victories in agrarian reform struggles for forests and against the asociaciones en participación contracts led to Durango being the scene of the first formally established CFEs in Mexico, among them the communities of San Esteban in 1964, La Ciudad in 1967 (with 13,975 hectares), Vencedores in 1968 (with 23,700 hectares), and San Bernardino Milpillas Chico, an indigenous Tepehuán community with 153,202 hectares, in 1969.

Peaceful protests by forest-based movements commonly found their causes promoted and supported by both a federal agency and the president, but more violent protests against abuses in some other states were not tolerated. In 1965 a brief armed uprising in Madera, Chihuahua, supporting land invasions and against abuses of Bosques de Chihuahua and other large landowners was quickly suppressed, but its occurrence began to focus presidential attention on the situation in that state (Henson 2019). Oaxaca also became the scene of grassroots protests when in 1964 the Sierra Sur community of Santiago Textitlán blocked roads, accusing the Forest Company of Oaxaca (Compañía Forestal de Oaxaca; CFO) of illegal logging and damaging the forest (UCEFO 1989). In 1967 in the Sierra Norte of Oaxaca, fifteen communities led by San Pablo Macuiltianguis formed the Union of Raw Material Supplying Towns to FAPATUX (Unión de Pueblos Abastecedores de Materia Prima a FAPATUX—Fábricas de Papel de Tuxtepec [Tuxtepec Paper Factories]) exercised their hold-up rights and refused to sign the logging contracts. They then launched a boycott for higher salaries, an increase in the stumpage fee, investment in roads, and scholarships for their children. A comunero in Macuiltianguis protested that "we seem like workers and not owners of the forest. That's why we have always had a rebellious position with respect to the enterprise since it carries away all our wealth and our sweat and doesn't leave us anything" (Alatorre Frenk 2000, 125). The *unión* persisted in its strike for six years before FAPATUX finally ceded to some community demands. But unlike Durango, the movement was primarily focused on labor issues and did not envision the option of managing the forests themselves. It would not be until the early 1980s that Oaxaca communities would join Durango in grasping the vision of managing their own CFEs. This transition has been described as a change in terrain (*cambio de terreno*) for peasant movements, from the historic demand for land to a new demand for the appropriation of the production process (Bartra 1991). Although Oaxaca has received more academic attention, it is clear that Durango and Chihuahua, the two largest forest states, were the pioneers in the formation of the CFE sector, with significant clusters of CFEs occurring by the end of the 1960s. However, it was the 1970s that brought the large-scale, state-directed strengthening of the existing CFEs and the creation of new CFEs, opening new spaces for responses by vigorous community collective action.

THE GREAT AWAKENING OF MEXICAN COMMUNITY FORESTRY 1971–1986

Sweeping initiatives to promote community forestry emerged during the presidency of Luis Echeverría Álvarez (1970–76). Echeverría attempted to address growing problems with the rural economy in general with state populist policies that included a major new Agrarian Reform Law in 1971 (Sanderson 1981). Echeverría focused on increasing agricultural production, tenure issues, peasant enterprises, and income inequality (Grindle 1977; Moguel and López Sierra 1990; Sanderson 1981), but his major impact on the forest sector has been little discussed. As noted by a government official at the time for forest communities, "The interest of President Luis Echeverría Alvarez, interpreted fully by the top officials of the Secretary of Agrarian Reform, is to train the campesinos so that they can be converted into owners of their enterprises" (Mendoza Medina 1976, 65). During a campaign visit to Durango in 1970, Echeverría promised support to the emerging CFE sector, and "upon [his] taking possession of the office, there began to immediately arrive resources from FONAFE to finance ejido industrialization" (Guerra Lizárraga 1991, 100). The reforms in the rural sector and in forests were driven by multiple factors, but in Chihuahua they were motivated by the disturbing memory (for the Mexican government) of the 1965 Ciudad Madera uprising and in Guerrero by an active and growing guerrilla insurgency in the early 1970s, abuses by logging companies being significant in both cases. Thus, the Echeverría presidency saw the launching of four different state policies that impacted forest communities: (1) the creation of timber parastatals known as Decentralized Public Organisms (Organismos Públicos Descentralizados; OPD), intended to exercise greater state control but also to serve as regional development agencies to provide community services and to organize and endow incipient CFEs, particularly in states marked by unrest such as Chihuahua and Guerrero (Ovando H. 1979); (2) the substantial expansion of the National Fund for Ejido Promotion (Fondo Nacional de Fomento Ejidal; FONAFE) to provide human, social, and physical capital to forest communities to establish their own CFEs. FONAFE had been created in 1959 to channel the deposited stumpage fees into investments in community productive capacity (though little of it reached the communities), but it now took on a substantially larger role (Moguel 1989); (3) a continued agrarian reform process of land and forest distributions to communities and the creation of new collective action spaces by requiring the organization of ejido unions; (4) the General Directorate for Forest Development (Dirección General de Desarrollo Forestal; DGDF), founded within the Forestry Subsecretariat of the Secretary of Agriculture in 1973 by forester León Jorge Castaños and directly authorized by President Echeverría (interview with L. J. Castaños, December 6, 2016), with an explicit vision of developing CFE capacity outside of the major concession areas.

THE TIMBER PARASTATALS (OPD)

The first large timber parastatal established was in Durango in 1968, Mexican Forest Products (Productos Forestales Mexicanos; PROFORMEX), though it did not begin to operate until 1973. After 1970 Echeverría vigorously expanded the model, establishing six more, including Forest Products of the Tarahumara (Productos Forestales de la Tarahumara; PROFOTARAH) in Chihuahua, Forest Promotor of Michoacán (Promotora Forestal de Michoacán; PROFORMICH), the Vicente Guerrero Forest Company (Forestal Vicente Guerrero; FOVIGRO) in Guerrero, and Industrialized Timbers of Quintana Roo (Maderas Industrializadas de Quintana Roo; MIQROO) (Enríquez Quintana 1976), merging regional development policy and forest policy. The largest of the parastatals was PROFOTARAH, in Chihuahua. Although many of the communities supported by the CCIT in the 1950s and 1960s had achieved varying degrees of self-management, logging abuses elsewhere in the Sierra Tarahumara were rampant in the early 1970s. Newspaper reports of some of these abuses stimulated Echeverría to meet in Mexico City with the indigenous organization the Supreme Council of the Tarahumara, and he subsequently ordered the formation of PROFOTARAH in August 1972 to try and curtail the abuses. Both a state logging company and a development agency, PROFOTARAH had the goal of developing community capacity to manage their forests, endowing many communities with physical capital in sawmills and logging roads and training for human and social capital, as well as clinics (interviews, Chihuahua, October 2015). PROFOTARAH soon displaced the CCIT, worked with the sixteen communities promoted by the latter, and went on to lay the groundwork for CFEs in an additional seventy-seven ejidos and comunidades. Social capital was further accumulated by organizing thirty of these CFEs into an intercommunity organization with a joint sawmill in Tomochi, Chihuahua (Boyer 2015).

Like the CCIT before it and with the power of the presidency behind it, PROFOTARAH began intervening in the contracts between communities and logging companies, and community members began to staff sawmills for the first time (Boyer 2015). In the 1970s PROFOTARAH appeared to have been a highly positive force in Chihuahua, making substantial investments in physical, social, and human capital that laid the foundations for the contemporary CFE sector (Estrada 2018). However, by the 1980s it had come to be seen as an oppressive bureaucracy that attempted to suppress traditional livestock grazing in the forest and to centralize sawing capacity away from community mills it had earlier supported (Boyer 2015). PROFORMEX in Durango, based in Santiago Papasquiaro, also played a positive development role in the first years—showing community members the value of their forests, providing physical capital in logging roads, and training in the basic tasks of industrial forestry—but in later years became more abusive in issues such as overharvesting forests and resisting the provision of communities with sawmills (interview with F. Salazar, February 8, 2019).

In Guerrero, FOVIGRO was established as part of a counterinsurgency campaign against a guerrilla movement headed by former schoolteacher Lucio Cabañas. The campaign combined "state terror" that swept up both guerrillas and innocent civilians in torture and disappearances with road construction, credits, and support for CFEs as a response to abuses by small-scale loggers that had created protests in the region for years (Aviña 2014; Bartra 2000). FOVIGRO was reported to be effective in the first years of its existence, providing support to forest communities in Guerrero to manage their own CFEs, before becoming increasingly politicized and inefficient in later years (interview with former high-level forest official, December 6, 2016). The performance of the parastatals was mixed, but in their first years they provided some of the first major injections of capitals for communities to eventually manage their own CFEs. As the parastatals became more oppressive in later years, some communities had to mobilize against them in order to maintain and expand organizational spaces and claim sawmills as their own, but frequently they also had allies in high places in these struggles, who intervened in support of community forestry (García-López 2013).

FONAFE

FONAFE, a unit of the Secretary of Agrarian Reform, collected the profits generated by logging; 70 percent went to FONAFE accounts in the community's name and the other 30 percent went directly to the community. As of 1971 it gained new powers to finance logging operations (and in 1976 underwent a change in acronym to FIFONAFE). In the beginning, FONAFE specialized in physical and human capital by financing sawmills and training in sawmill operation, logging, and marketing, and its explicit goal was to create "ejido forest enterprises" (Enríquez Quintana 1976). The fund established training centers in Chihuahua and Durango with a staff of thirty-three professionals and a trust fund (*fideicomiso*) to operate the National Program in Training Labor for the Silvicultural Industry whose purpose was implementation of "a massive type of training" (Mendoza Medina 1976, 52). It provided training in some twenty-two different skills, including saw sharpening, paperwork documenting, logging supervision, sawmill administration, machine operators, diesel mechanics, reforestation, seedling nurseries, carpenters, lumber classification, and more, a significant infusion of human capital (Mendoza Medina 1976). It eventually promoted 135 CFEs in fifteen states, with eighty-seven of them being funded for sawmills, a major endowment of physical capital. Ninety-two of them, or 68 percent, were in Durango and Chihuahua, and a second significant cluster of twenty ejido forest enterprises emerged in the tropical states of Campeche and Quintana Roo (Enríquez Quintana 1976). The physical reminders of these endowments are still quite visible today. In 2014 I visited the community of Cusárare in Chihuahua to find community members still operating a very antiquated but functioning sawmill endowed by FONAFE in the early 1970s. In the community of Los Altares in Durango in 2019, I saw a plaque commemorating a visit by Echeverría in 1976 to inaugurate a sawmill there. This rapid formation of CFEs had a positive impact on forest production, with production in Durango nearly tripling from 1970 to 1980; in 1975, ejido enterprises constituted 21 percent of total national production (Guerra Lizárraga 1991). In Michoacán, FONAFE established a resin-processing plant with sixty-two supplier communities (Enríquez Quintana 1976). In Quintana Roo, the fund launched logging of tropical hardwoods for railroad ties in Quintana Roo and Campeche, and ejido unions were established in both states in the early 1970s. Also in Quintana Roo in the early 1970s, the earliest community-located, though not community-administered, sawmills were established in Nicolás Romero and Cafetal-Limones. Santa Cruz Tanaco in Michoacán, now defunct but considered one of the model CFEs in the 1970s, was established with FONAFE funds in 1973 (Vázquez León 1992). Although some of these FONAFE efforts also failed in the short run, these failures sometimes spurred later successes: an effort to establish an entrepreneurial alliance called IXCAJIT (Ixtlán-Capulalpan-Xiacui-Trinidad) in Oaxaca in 1974 quickly collapsed, but the important CFE of Ixtlán de Juárez emerged out of it with new government-endowed physical capital.

FONAFE was encharged with organizing the CFEs as exclusive suppliers to the parastatals and has been correctly accused of paternalism. However, at the same time, the state and private forest industries accused the top-down recurrent reformers of being "socialist" (Castaños Martínez 2015, 92). Their efforts were certainly paternalistic and top down. Moreover, prices for timber sold to the parastatals were set administratively, and thus communities were not directly engaged with market prices. Nonetheless, it is hard to see how communities composed almost entirely of subsistence corn farmers with low levels of education and no prior experience in industrial logging could have made the transition to managing their own CFEs any other way than through this top-down infusion of human, physical, and social capital, along with titles to the natural capital of the forests, driven by reformers accused of being socialists. There was also a genuine vision of devolving management responsibilities among the state reformers: "The campesinos participate in the decision-making of the Enterprise, participate in the benefits derived from the industrialization and marketing of their forests, the income is not concentrated in a few hands, the available labor is occupied, the forest resources are technically exploited, and for their protection and development since the forest is seen by the campesinos as a permanent source of raw materials supply for the enterprises on the property where they and their children work" (Enríquez Quintana 1976, 72). According to FONAFE's director, the fund's plan was that "in a period no more than ten years, the campesino will take charge of the complete administration of the forest enterprises" (Garzcón Mercado 1975).

AGRARIAN REFORMS

Land and forest redistribution actually declined from 23.7 million hectares under the previous administration of Díaz Ordaz to 12.8 million hectares under Echeverría. However, it was still the third-highest redistribution historically (Núñez Rodríguez n.d.) and included some major distributions of forest land. Explicitly identifying it as a response to the Ciudad Madera uprising of 1965, Echeverría created in 1972 the enormous Ejido El Largo out of an existing concession to Bosques de Chihuahua (Henson 2019; Moguel and López Sierra 1990) and ordered FONAFE to support the process of creating a CFE. El Largo, with a vast 265,111 hectares, immediately became Mexico's largest ejido and largest CFE, with a twenty-year contract requiring it to continue to sell to Bosques de Chihuahua. Nonetheless, over time, the community members gradually learned how to operate a CFE and achieved greater autonomy that would eventually make it "likely the largest CFE in the world" (Hodgdon and Estrada Murrieta 2015, 3). The example of El Largo also inspired a rebellion against existing logging contracts in another enormous Chihuahua ejido, Chinatú, with over 150,000 hectares, resulting in the beginnings of greater autonomy in their operations (Boyer 2015). Also in the 1970s El Balcón, in the municipality of Técpan de Galeana, Guerrero, received its major forestlands endowment of nearly 20,000 hectares (Torres Rojo, Guevara-Sanginés, and Bray 2005). It was no accident that Técpan de Galeana was one of the areas of operation of the Cabañas insurgency, and he was actually killed in the municipality by the military in December 1974 (Aviña 2014). Given the choice between joining the guerrillas and the opportunity to start organizing a CFE on nearly 20,000 hectares of rich forestlands, El Balcón chose the latter (see chapter 4 for more detail). In addition to the forest redistributions, new collective action initiatives were in fact *required* from above by a new agrarian reform law.

The 1971 Agrarian Reform Law sought to broaden ejido economic options by promoting ejido unions or "forest associations" (García-López 2013; Moguel 1989) that could jointly market products, buy inputs, and channel state support. The ejido unions were subordinate to government agencies such as the Rural

Credit Bank and the Secretary of Agrarian Reform (Merino Pérez 2004). But in many cases these efforts also created linking and bonding social capital at the community and intercommunity level, providing an important foundation for the emergence of CFEs and enduring intercommunity organizations (Torres Rojo and Graf Montero 2015). The ejido unions, new players operating under new rules of the game, constituted a massive top-down organizing effort, and as of 1976 there were twenty-five forest ejido unions with 611 ejidos in thirteen states (Mendoza Medina 1976). Most did not survive in that organizational form but a few remain viable and important today, such as the Union of Forest Ejidos and Communities Emiliano Zapata (Unión de Ejidos y Comunidades Forestales Emiliano Zapata; UNECOFAEZ) in Durango and the Hermenegildo Galeana Ejido Union (Unión de Ejidos Hermenegildo Galeana; UEHG) in Guerrero. As has been noted for agricultural ejido unions, "Even if it was not the original purpose, creating these organizations augmented the negotiating capacity of the ejido, giving birth to a new generation of peasant leaders" (de Janvry et al. 2001, 56).

THE DGDF

FONAFE promoted state-led community forestry as suppliers to the concessioned state enterprises. In contrast, the DGDF marked a state-led reform effort at developing CFEs outside of the concession areas. It was explicitly conceived as "an agent of change" and had the view, radical even for today, that forestry was principally "human development" (Castaños Martínez 2015, 90–93). This policy shift, heralded in 1973 by the elaboration of the first National Program of Forestry Development, was to be implemented by the DGDF. Under the leadership of León Jorge Castaños, it launched an outpouring of studies and proposals. Some sixty-nine studies were produced between 1973 and 1982 (Calva et al. 1989), with one of the most important being the study that justified lifting the logging bans still in place in many states, marking an end to the use of that forest policy, and allowing the DGDF to operate in regions that previously had bans. The DGDF developed a remarkable strategic vision of what would later be called multilevel governance and of the communities as "collective action common properties" with continued state regulation (Castaños Martínez 2015, 99). Of the DGDF team in Oaxaca (see later in this section), it was noted that their members were "fierce critics of the system of forest concessions and of the corruption present at the time in forest administration. They believed in socialism in general, and in particular in the capacity of the peasant communities to 'appropriate the productive process' and construct modern communal enterprises that could attack poverty and drive self-managed development" (Garibay Orozco 2008, 191).

After initial efforts in the state of Tlaxcala, the DGDF had its first successful initiative in the neighboring Chignahuapan-Zacatlán area in Puebla State, which had larger forest masses but also a process of intense deforestation. These states had only small-scale contractors working illegally, so there were few political risks and in January 1975 the DGDF was able to get the logging ban lifted in Puebla. The DGDF was also important in the organizational processes in Quintana Roo and eventually worked in nine states (see later in this chapter for more detail on the DGDF in Puebla, Oaxaca, and Quintana Roo). The DGDF also began to introduce a new silvicultural practice, the Silvicultural Development Method (Método de Desarrollo Silvícola; MDS), first in 1976 in the UIEF Atenquique in Jalisco and later in Puebla and other states, as an alternative to the selective cutting practices of the Mexican Method of Forest Zoning (Método Mexicano de Ordenación de Montes; MMOM). The ecological significance of this change is explored further in chapter 7.

THE MATURING OF REFORMS AND COMMUNITY MOBILIZATION 1976–1982

The end of the Echeverría presidency and the ensuing presidency of José López Portillo (1976–82) marked a partial end to the wave of forest reforms, including a dramatic reduction in the distribution of new forest lands, the end of FON-AFE investments, and a decline in positive development influences from the parastatals, though promotion of ejido unions continued (Moguel and López Sierra 1990). Nonetheless, foundations had been laid and capital investments made, and the forest reform process matured in these years. In particular, the DGDF was able to persist in its community organizing, working in areas where there were few powerful economic actors. Political cover was provided by the fact that in 1978 Cuauhtémoc Cárdenas, the son of the revered former president, became the forestry subsecretary. Cárdenas assumed the mantle of the forest reformers, affirming the goal that the owners of the forests should work them directly and receive the benefits from them (SARH 1979). Nearly four years of intensive extension work by the DGDF's Plan Puebla initiative resulted in the conception and realization of a new community enterprise organizational form, with associated institutions, the Raw Materials Ejido Production Units (Unidades Ejidales Productores de Materias Primas; UPMP). The DGDF also developed the concept of "socioproduction" to describe its strategy: "Socioproduction is thus named because it is based on the idea that social necessities are resolved by production which in turn, promotes the development of society. It is also seen as a vehicle for promoting forestry culture" (Aguilar Espinoza 1990, 12). By 1978, the DGDF had promoted an intercommunity forest association, Bosques y Maderas de Chignahuapan-Zacatlán (Forests and Timbers of Chignahuapan-Zacatlán; BOMACHIZA), and CFEs with logging permits had been organized in fifty-two ejidos as well as 253 small private forests, while getting from 50–300 percent better price per cubic meter than in unorganized communities (AMPF 1979; SARH 1979). As it achieved success in Puebla, the DGDF received more resources and in 1978 began expanding into the Cofre de Perote and Huayacocotla regions of Veracruz, where concessions to parastatals also did not exist, in 1980 into Nuevo León, and in 1982 into Quintana Roo. In 1981 a new intercommunity forest association was formed in Huayacocotla, though efforts in the Cofre de Perote ran into a new logging ban pushed by environmentalists. But despite the new ban, at least one small ejido promoted by the DGDF, Rosario-Xico, established a CFE and continued on until today as a successful forestry ejido (P. Geréz, personal communication, June 10, 2019). This period was characterized later by a DGDF staff member as the time when "at last, we began to speak of a country of silviculturists, of forming social forest enterprises and, in sum, of democratizing the process of forest production" (González Martínez 1992, 5).

Whereas the FONAFE initiatives focused principally on technical aspects of forestry and accounting in building human capital and endowing physical capital, the DGDF explicitly focused on community enterprise governance as well, creating new organizational models for how an UPMPF or CFE could be inserted in the existing governance institutions of the communities. DGDF officials produced various organizational charts with governance options, as shown in figure 4, providing models from above that reduced the transaction costs of the communities. The "legal, political, and administrative level" (upper left quadrant) was established in agrarian law. The "silvicultural productive level" was the DGDF proposal for different organizational models for structuring the CFE. Varying occupational roles elected by the Assembly and placed under the supervision of the Comisariado included a coordinator, *jefe de monte* (forest

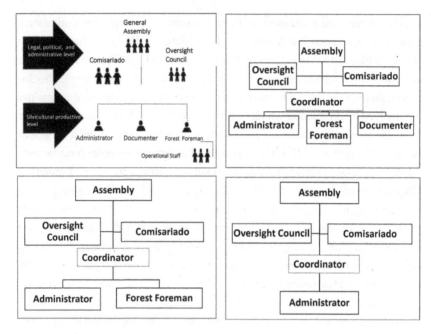

FIGURE 4. Organizational variants for forest material production units (DGDF). Chart adapted and translated from Castaños Martínez (2015, 100–101).

foreman), administrator, and the paperwork documenter (*documentador*), among others.

This was the first documented conscious creation of a new governance arena of collective choice in forest enterprise management whereby communities could assert autonomy over enterprise governance, with organizational models provided from above but which they could adapt to their needs. In human capital formation, the DGDF developed strategies of "promotion and rural communication, adapted to the characteristics of the peasants in language, preferences, and levels of comprehension, with the purpose of achieving their involvement in their own productive units" (Castaños Martínez 2015, 100). It had an entire unit dedicated to communication and produced eleven training manuals in comic book form. By 1984 it came to have a national staff of some 300 people, only 15 in the central offices and the rest decentralized in the states (interview with L. J. Castaños, December 6, 2016). It also developed strategies of involving state and federal government, the private sector, and communities in what today would be called multistakeholder collaborations.

GRASSROOTS MOVEMENTS IN THE 1970S AND 1980S; CONTINUED ROLE OF THE DGDF

While the 1970s were most notable for the panoply of top-down state-led reform efforts, the grassroots protests in Durango in the 1960s reemerged in the 1970s and early 1980s in Durango and Oaxaca. In Durango new and more abusive management in PROFORMEX and close advising support from Agrarian Reform stimulated twenty communities to form the ejido union UNECO-FAEZ and demand more employment by locating sawmills in communities. Agrarian Reform arranged for a direct meeting with President Echeverría, and he intervened directly with PROFORMEX to order them to establish more sawmills in the communities, another case of a Mexican president and a federal agency acting in direct support of the emergence of the CFE sector against economic and political interests at the state and federal level (interview with R. Vidaña, February 6, 2019). As a result of this intervention, there were seven sawmills in UNECOFAEZ communities by 1980. I noted earlier seeing the plaque on a still-functioning sawmill endowed by Echeverría in 1976 in Los Altares, Durango. The sawmills were still owned by the parastatal PROFORMEX, but it now provided employment and invested in human capital by teaching sawmill operations to community members. UNECOFAEZ remained tied as suppliers to PROFORMEX as late as 1987, when it organized a tense roadblock over control of the logging documentation. This conflict led finally to the union's complete independence and full direct engagement with market price incentives in that year (Garcia-López 2013; Taylor 2000). The union members also showed considerable political acumen, by tying themselves close to the then-ruling party and engagement in the presidential campaign of 1988, assuring continued high-level political support for their efforts (G. Chapela y Mendoza 1998). This engagement paid off in 1989, when President Salinas de Gortari endowed UNECOFAEZ with a plywood factory and other timber-processing businesses with the dissolution of PROFORMEX, again siding with communities against states' economic interests (see chapter 6 for more detail).

In Oaxaca, grassroots protests began to occur in the early 1970s. The community of Santiago Textitlán, starting in 1973, carried out a production strike for three years, launching its own CFE by 1978, though still under a legal obligation to sell to a private concessionaire. It functioned only three years, but was resuscitated by a DGDF team in 1984 (ASETECO 2003). In 1976 Pueblos Mancomunados carried out a stoppage against the private company Maderas de Oaxaca, seizing equipment in protest, and went on to form its CFE in 1977 (UCEFO 1991). In 1979 San Pedro el Alto seized machinery belonging to a concessionaire and exercised its hold-up rights to stop logging until 1983, when it was also helped to establish its own CFE by the DGDF (ASETECO 2003; see chapter 4 for more detail). Unrest in the forests of Oaxaca reached a boiling point in the early 1980s, when the Organization in Defense of Natural Resources and Social Development of the Sierra Juárez (Organización en Defensa de los Recursos Naturales y Desarrollo Social de la Sierra de Juárez; ODRENASIJ), composed of twenty-seven communities, emerged fighting against the renewal of a twenty-five-year concession to the parastatal FAPATUX, due to end in 1982, in a struggle for direct market access and the right to form their own CFEs (Abardía Moros and Solano 1995). In contrast to the direct state support in Durango, this effort was supported by a coalition of student activists with training in forestry and community development and was apparently the only place in Mexico where student activists had a major impact on community forest organization. The students drew on a variety of inspirations—including Maoist, socialist, humanist, and Christian—and "the magic words . . . were participation of the base, self-management, participatory democracy, horizontality, social appropriation of the productive processes, control of the resources by the owners and possessors" (Alatorre Frenk 2000, 52). Among other things, this was clearly an important period of construction of social capital, where a major function of the advisors was getting previously isolated and frequently warring communities to talk to each other about common interests.

The movement against concession renewal caught fire in November 1982, with a presidential decree from López Portillo that renewed concessions over the entire state of Oaxaca, one of the few documented examples of presidential power used against the community forestry sector. ODRENASIJ and its advisors launched a campaign against the concession that included the strategy of an *amparo*, a legal mechanism that allows Mexican citizens to protect themselves from negative actions of the government. They also created a national publicity campaign about their struggle and enforced a complete production strike. In early 1983 the amparo was upheld, with the Mexican legal system supporting community forestry. When President Miguel de la Madrid entered office in December 1982, one of his first actions was to definitively cancel the concession, a reversion to this form of direct presidential support for the community forestry sector. However, by the end of 1983 ODRENASIJ had collapsed. With the victory over the concession won, the communities had free access to markets, and the hard part of building their individual CFEs had begun (Abardía Moros and Solano 1995; Bray 1991).

The student-accompanied struggles in Sierra Norte also helped to consolidate the first forestry nongovernmental organizations (NGOs) in Mexico. The Environmental Studies Group (Grupo de Estudios Ambientales; GEA), possibly the first environmental NGO in Mexico, was founded in the 1970s, with some founders being staff members of the DGDF. The group was active in the Sierra Norte in the early 1980s (Aguilar et al. 1990). Rural Studies and Advising (Estudios Rurales y Asesoría; ERA) was founded by student activists Francisco Chapela, Yolanda Lara, Sergio Madrid, and Xóchitl Ramírez, all of whom had advised the ODRENASIJ movement and the first efforts at organizing CFEs. ERA became the first forest NGO to receive international foundation support from the Ford Foundation, in 1985, and it went on to organize another early intercommunity organization, the Zapotec-Chinantec Union (Unión Zapoteca-Chinanteca; UZACHI), one of the few organized by an NGO rather than the state.

With the victory of the anticoncession movement, DGDF state reformers also entered Oaxaca in 1983. The DGDF first took up residence for a year in the community of Latuvi in Pueblos Mancomunados and began an organizational process to wrest control of a "community" sawmill that had been taken over in an elite capture since 1976. The DGDF team, led by Rodolfo López Arzola, helped the community establish a genuine community sawmill in 1983. Using the model of the UPMPF but focusing more strongly on business aspects, three members of the DGDF team took up residence in Santa Catarina Ixtepeji for a year in 1983 to launch its CFE. The intensity of human capital formation was captured by a member of the DGDF team:

In Ixtepeji we dedicated ourselves to training in all aspects, both the orientation for organizing the work in the forest and in the enterprise, and the profiles for the roles, the nominations, etc., everything that an enterprise brought with it. Then we dedicated ourselves to training the administrators. It was the first experience of a campesino in that office. We stuck to that man, that first administrator, who was an older man, Don Beto (who had not even finished primary school), up to the point where he became very confident . . . [H]e was a person of much respect and much presence in his community, a very honest person. He was there three years, and when he finished he had a series of skills that surprised himself and commented "the truth is that I don't know why those little accountants [*contadorcitos*] are paid so much for this work." (ASETECO 2002, 63)

The DGDF team also helped Ixtepeji develop written community statutes or rules that included the rights and obligations of the community and the new organizational player of the CFE. The DGDF team in Oaxaca, which eventually grew to twenty-three people and began working in various parts of the state, had the strategy of placing a community organizer full time in each community where they were working. However, there were also intense bureaucratic fights, with important elements of the "forest old guard" in the government resisting these efforts, another example of the intra-government bureaucratic struggles that accompanied the emergence of community forestry. Eventually López Arzola left the government and helped form the first independent forest association in Oaxaca, the Union of Forest Ejidos and Communities of Oaxaca (Unión de Ejidos y Comunidades Forestales de Oaxaca; UCEFO) in 1985 and later the forest NGO ASETECO (ASETECO 2002).

While the tumultuous events in Oaxaca were producing significant numbers of new CFEs by both bottom-up grassroots mobilizations and top-down state reforms, another state-led reform was occurring in the tropical state of Quintana Roo. The Plan Piloto Forestal (Forest Pilot Plan; PPF) emerged in 1983, constituted by a unique confluence of Quintana Roo, federal, and international actors. The Acuerdo México-Alemania (Mexico-Germany Agreement; AMA) brought in substantial financial and technical assistance from the German foreign aid agency German Corporation for International Cooperation (*Gesellschaft für Technische Zusammenarbeit*; GTZ), with the DGDF providing salaries and political support. Further political support came from a young governor of Quintana Roo, Pedro Joaquín Coldwell, who was developing the megaresort of Cancún and was convinced that community forest management would ensure tourists were not greeted by a deforested state. At the same time, communities were growing restive around the end of a twenty-nine-year logging concession granted to the parastatal MIQRO, though with little in the way of the grassroots mobilizations of Oaxaca (Bray 2004a; Bray et al. 1993; Galletti 1998; Merino 1997; Taylor and Zabin 2000).

In a series of institutional and forest management innovations, the PPF helped the communities in declaring permanent forest areas (PFAs), or áreas forestales permanentes, conducted participatory forest inventories, established CFEs, and erected intercommunity forest associations that served as the channels for technical assistance, donor support, and negotiations with government agencies. From 1985 to 1989, 502,166 hectares were placed in PFAs in southern and central Quintana Roo by decisions in ejido assemblies. The PFAs provided for selective logging under management plans and created a fixed internal agricultural frontier in each ejido, forcing slash-and-burn agriculture to operate within more confined areas. In addition, two forest associations were founded in 1986 by initiative of the PPF organizers, the Society of Forest Production Ejidos of Quintana Roo (Sociedad de Ejidos Productores Forestales de Quintana Roo; SPFEQR), with ten ejidos (Wilshusen 2009, 2010), and the (Organization of Forest Production Ejidos of the Mayan Zone; Organización de Ejidos Productores Forestales de la Zona Maya; OEPFZM), with thirteen ejidos (Bray 2004b; Bray et al. 1993; Merino et al. 1997; Merino Pérez 2004). The SPFEQR, with some exceptions, was composed mostly of more recent colonist communities dominated by mestizos. The OEPFZM in the Mayan Zone of central Quintana Roo faced a different situation, comprising much poorer, heavily monolingual Mayans with a tradition of fierce resistance as descendants of the protagonists of the Caste War of Yucatán (Dumond 1997). Because of limited resources and the difficulty of the challenge in the Mayan Zone, the PPF made a conscious decision at an early stage to drastically reduce support to that team, leaving it to struggle to find its own resources. The founding of the OEPFZM also emerged from a deeper social struggle than in the South. The Mayan ejidos of the area had been organized by FONAFE into unions in the mid-1970s to produce railroad ties, but they had quickly fallen into corruption. The OEPFZM struggled against those ejido unions, which finally disappeared when the OEPFZM spearheaded the direct marketing of railroad ties and the development of forest management plans (Bray 2004a, 2004b; Bray et al. 1993; Bray et al. 2004). As a representative with the Inter-American Foundation, I first visited the OEPFZM in 1991 and was able to channel the organization its first external funding. The OEPFZM has survived up to today, still providing multiple support services to the Mayan communities of central Quintana Roo.

THE CONSOLIDATION OF COMMUNITY FORESTRY (1983–1988)

Both state and grassroots initiatives were greatly aided by the fact that León Jorge Castaños was named as forest subsecretary in 1983, bringing a reformist vision to the highest level of the forestry agency. However, budget cuts began to impact the agency, and by 1985 the forestry subsecretary was bureaucratically downgraded to a National Forest Commission, though the DGDF continued as a unit through 1988 (interview with L. J. Castaños, December 6, 2016). Despite bureaucratic setbacks, Castaños was able to push through a new forestry law that contained many of the elements promoted by the DGDF over the years. The watershed 1986 Forest Law, characterized as "peasant-oriented and environmentalist" (G. Chapela y Mendoza 1996, 355), ended all private concessions and initiated a process of dismantling the parastatals, allowing CFEs across the country to engage directly with prevailing market prices for timber; required more detailed and environmentally sensitive studies for logging permits; established Units of Conservation and Forest Development (Unidades de Conservación y Desarrollo Forestal; UCODEFO) to serve as regional forest technical service providers; and authorized communities directly or through forest associations to operate their own forest technical services (*servicios técnicos forestales*). Forest technical services refer to management plans that must be drawn up by professional foresters for logging permits to be issued. Historically, these services had been monopolized by the government and associations of forest engineers in the UIEFs and other organizations (Bray, Antinori, and Torres-Rojo 2006). However, between 1988 and 1992, some forty technical service concessions were given to communities, creating new incentives for collective action around CFEs.

Thus, 1986 was the high-water mark of the community forest sector in Mexico's first period of maturation; these gains resulted from reforms going back to the 1950s, driven both by top-down reformers and grassroots mobilization, though the latter were most relevant in Durango and Oaxaca, while nationally reforms from above were more influential and crucial. The CCIT, PROFOTARAH and other parastatals, FONAFE, and the DGDF all constituted major and relatively well-funded long-term reforms from above that together stretched over more than a decade and a half. The NGO effort that supported the creation of UZACHI in Oaxaca was also a long-term effort, lasting some ten years. These public and civil society efforts provided intensive support to incipient CFEs for sustained periods of time. A little-noticed feature of all these efforts was that they had major extension components in addition to endowments of natural and physical capital; that is, they had government forest extension agents who spent substantial time in communities training and promoting social and human capital. Before this period, forest service employees were almost entirely urban based, processing paperwork: "the Forest Service has a city-based personnel, living permanently in the national and state capitals, and . . . lacks rural personnel that stays in constant contact with the forested areas" (Hinojosa Ortiz 1958, qtd. in Mathews 2011, 50).

In contrast, an official with PROFOTARAH in the 1970s noted, "the most important part was that the operational people were in the field—they weren't in Chihuahua City. They lived in camps in the forest; they worked twenty-two days, and then had eight days of rest. They were constantly accompanying the communities in the cultivation of the forest in every respect. That was the success" (interview with A. Quiñones, December 18, 2013). The DGDF in Puebla had a similar strategy, "Professionals and technicians lived and stayed in their work area, they had permanent contact with the people and their problems; technical tourism was eliminated and the technical and professional personnel in state headquarters and representations were reduced to a minimum" (Castaños Martínez 2015, 107). In Puebla ten years were needed for the first stage of CFE vertical integration: producing timber ready for transport at the skid trail. FONAFE only existed as such for six years, but the CCIT operated continuously for over fifteen years and PROFOTARAH nearly twenty, though the period of effective extension was apparently confined to the 1970s. The DGDF organizers in Oaxaca lived in communities for a year or more at a time, and the PPF in Quintana Roo was a constant presence for up to two decades in building human and social capital and providing physical capital in the form of sawmills; many of the advisors have continued on until the present in different organizational forms.

So how many CFEs were operating by 1988, as a historical marker for the growth of the CFE sector? It is hard to say with precision, but available numbers provide some idea. In the mid-1970s, there were tallied 286 ejidos with contratos de asociación, 453 with contratos de compra-venta, and 257 ejido forest enterprises (Mendoza Medina 1976). Making the possibly risky assumption that the majority of these CFEs continued logging and eventually became more autonomous, and including the total of 310 ejidos and communities promoted by the DGDF (Castaños Martínez 2015, 114), one can estimate that there may have been over 1,300 CFEs operating to one degree or another by 1988. Therefore, the Mexican community forestry sector likely substantially had come into being by that date, with subsequent years being a maturation and broadening of that foundation. The coupled SES of Mexican community forestry had emerged, supported by the institutional framework of Mexico and the collective action of the CFEs, with millions of hectares of forest now under generally more careful stewardship by communities in their own forests.

This concentrated, relatively well-financed, mostly top-down reform effort in institutional and capital supply through forest extension is what it took to make this great leap forward in Mexican community forestry. There were important grassroots mobilizations that forced open further collective action spaces that could develop community-based economic alternatives in Durango and Oaxaca, but even these had significant state support and timely presidential interventions. However, in most other states, the process was less about protest and more about communities vigorously seizing the opportunities that came from above to get into the timber business. A DGDF staff member noted, "The people were farmers and cattlemen, not silviculturists" (ASETECO 2002, 45). They were not only not silviculturists; they were also not chainsaw operators, skidder operators, logging truck drivers, bookkeepers, or administrators of vertically integrated market-oriented enterprises. For their forest industries to become fully community owned and operated was truly remarkable, and with the disappearance of the parastatals in the late 1980s the communities became fully engaged with market price incentives for collective action, a historic transition from state-led forestry to community-led forestry. This historic process of state reform and community response converted subsistence corn farmers with minimal grade school educations into managers of industrial forestry processes, constituting an SES that would become resilient to climate change and political and economic shocks. Since this formative period, the community forest sector has now had thirty years to mature and expand, and that process that will be covered in chapter 3.

CHAPTER 3

THE CONSOLIDATION OF COMMUNITY-LED COMMUNITY FORESTRY

1988–2018

T he 1986 watershed law and the end of the presidential period of Miguel de la Madrid in 1988 marked the termination of the first phase of the historic arc of Mexican community forestry. The period displayed the emergence of a distinct SES driven by state policy; a facilitative political regime, which provided institutions and organizational models from above and the five capitals; and increasing integration into open markets—all combined with a vigorous community response, ranging from social movements to the details of learning a business. The second phase, covering the thirty-year period from 1988 to 2018, is one of consolidation and maturing of the existing tendencies and has had ups and downs. This phase includes an apparent modest continued growth of the community forestry sector, mostly consistent state support for the sector, and strengthening of territorial zoning practices that controlled deforestation, conserved biodiversity, and provided environmental services. Figure 5 serves as a reference for the major programs and events by presidential period.

The presidency of Carlos Salinas de Gortari (1988–94) was marked by a wave of new legislation and policies that while exhibiting indifference and even hostility toward community forestry would also implement reforms that would end up having major impacts on the forest sector. These impacts are particularly evident in the devolution of an almost full bundle of property rights over forests, cementing the foundations for the counter-hegemonic alternative of

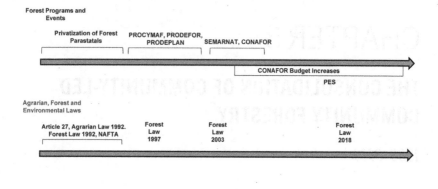

Recent Mexican Forest Policy Timeline (1988–2018)

Forest Programs and
    Events

Privatization of Forest   PROCYMAF, PRODEFOR,
    Parastatals   PRODEPLAN   SEMARNAT, CONAFOR

                                                CONAFOR Budget Increases
                                                    PES

Agrarian, Forest and
Environmental Laws

Article 27, Agrarian Law 1992.   Forest   Forest   Forest
    Forest Law 1992, NAFTA   Law   Law   Law
                                   1997   2003   2018

Presidential Periods

1988 – 1994   1994 – 2000   2000 – 2006   2006 – 2012   2012 – 2018   2018 – 2024

Salinas de   Zedillo Ponce   Fox   Calderón   Peña   López
Gortari   de León   Quezada   Hinojosa   Nieto   Obrador

FIGURE 5. Mexican forest policy timeline (1988–2018). Chart adapted from Mathews
(2006).

community enterprises that had begun with the agrarian reform of the 1930s.
Beginning in 1991, Salinas de Gortari launched an aggressive program of privatization of state enterprises and constitutional reforms. His goal was to introduce market forces into the countryside by creating pathways for ejidos, but
not comunidades, to privatize themselves in order to create a land market in
rural Mexico. However, significant environmental protection for forests in the
legislation provided powerful disincentives for dissolving an ejido that had any
significant forest resources.

THE SALINAS DE GORTARI REFORMS

The Salinas de Gortari period included at least five major policy initiatives that
impacted community forestry, the privatization of state enterprises, including
the timber parastatals, the modification of Article 27 of the Mexican Constitution, the 1992 Agrarian Law, the 1992 Forestry Law, and the North American
Free Trade Agreement (NAFTA).

THE PRIVATIZATION OF STATE ENTERPRISES

The privatization of the Echeverría-era timber parastatals definitively freed forest communities from the requirement to sell to them, as an unintended but positive consequence. As discussed in chapter 2, the burgeoning CFE sector initially emerged, to a significant degree, tied to these parastatals in a monopsony, a market where there is only one buyer. The privatization was the final step in dismantling the state tutelage of CFEs, a process that had been started with the 1986 Forest Law. The selling off of assets of the parastatals in the late 1980s and early 1990s, usually to CFEs and ejido unions, definitively engaged the CFEs with prevailing market prices for timber. As we will see in chapter 6, it is a crucial factor for the relative success of CFEs that these prices have been relatively high and relatively stable for many decades. For the first years, the CFEs frequently ended up selling to the same buyers for practical reasons, but eventually broadened their portfolio of buyers in more competitive markets (interview with J. M. Torres Rojo, February 12, 2017). The transfer of parastatal assets to CFEs and their associations provided major new physical capital in a number of cases. For example, the assets of PROFOTARAH were transferred to three ejido unions, after an initial failed effort to sell to a large construction company. In Durango, despite opposition from political and economic elites tied to timber, the state sold UNECOFAEZ the plywood factory that had belonged to the parastatal PROFORMEX. However, the factory's deteriorated condition meant they needed additional support to make it operational, which came in the form of low-interest state loans (Taylor 2000). Individual communities in Durango were also able to buy the sawmills on their lands from PROFORMEX, substantial transfers of physical capital.

THE REFORM OF ARTICLE 27 OF THE MEXICAN CONSTITUTION

The privatization of the timber parastatals opened up new economic spaces for CFEs at the same time that the second major initiative, the reform of Article 27, gave them strengthened rights over their forests. The reform of Article 27 and the accompanying 1992 Agrarian Law were an attempt to "recraft property rights" (Muñoz-Piña, De Janvry, and Sadoulet 2003), by opening up rural land markets. The 1917 Constitution had given communities long-term, stable usufruct rights; a degree of territorial control; a regime of local political and territorial governance institutions; and a multiscale governance framework. The agrarian reform had become a deeply rooted aspect of rural culture in Mexico (Bray, Antinori, and Torres-Rojo 2006). The reform sought to substantially reduce the common property sector in Mexico by allowing ejidos to have their land surveyed, parcel the lands into private landholdings, and dissolve the ejido by majority vote. Comunidades could not dissolve themselves, providing protection for indigenous territories, but could enter into joint ventures with private enterprise. However, the reforms also strengthened community rights over the common property by removing references to land belonging "originally" to the state. It essentially defined ejidos as a form of private property, a corporate communal property. This policy included some continued regulation of internal affairs by the state, though greatly reduced from previous periods, and has been characterized as a transition from a state-led common property regime to a community-led common property regime (de Janvry et al. 2001). Similarly, the period from the 1986 Forest Law through the Salinas de Gortari period constituted a transition from state-led community forestry to community-led community forestry.

THE 1992 AGRARIAN LAW

The devolution of greater property rights to the territory also devolved clearer rights over forests, with two important exceptions. First, ejidos could not legally parcel the forests (Article 59) and if an ejido voted to dissolve itself, its forests would become property of the nation (Article 29), a powerful disincentive for dissolution for communities with commercial forests. Second, although the communities had a nearly full bundle of rights for the trees, timber, and NTFPs, extraction would continue to be regulated by forest and environmental laws. The new agrarian law also impacted the institutions and organization of CFEs, since groups of individual ejidatarios now had the right to organize subcommunal enterprises or coalitions (newly emergent players known as "work groups") under a legal figure called "Rural Production Societies." This change allowed the work groups to access the common property resource when authorized and regulated by the community Assembly (Article 111) (Wilshusen 2005); see chapter 5 for more discussion of work groups. The Agrarian Law of 1992 also created a new property category of small forest property (*pequeña propiedad forestal*), with a limit of 800 hectares; ended the agrarian reform and land and forest distribution to encourage private-sector investments; legalized business alliances between ejidos and comunidades and private corporations; and created mechanisms for development of plantations of up to 20,000 hectares (Tellez Kuenzler 1994). Concerns frequently expressed were that these agrarian reforms would lead to perverse incentives to deforest, divide, and sell, but this does not seem to have happened on a significant scale. However, parceling of forests among ejidatarios, creating de facto but illegal private forest properties, has been a reportedly large but unquantified phenomenon.

THE 1992 FOREST LAW

The 1992 Forest Law was primarily focused on encouraging forest plantations, made little distinction between natural forests and plantations, made little reference to community forestry, and displayed a striking indifference to the sector that had been maturing over the previous fifteen years. The law also deregulated the transportation and processing of wood products, except for the origin of logs (World Bank 1995). This deregulation reportedly opened the door to a surge in illegal logging. For example, the expansion of the total number of sawmills in Chihuahua from 1993 to 1998 reportedly went from 108 to 309 as a result of illegal logging and the elimination of regulations on sawmills under the law (Guerrero, Reed, and Vegter 2000).

THE NORTH AMERICAN FREE TRADE AGREEMENT

Passed in 1993, NAFTA was feared to be a devastating blow to the forest sector in general, since in that year prices per board foot of wood were considerably higher in Mexico than in the United States. It was feared that many small community producers would go under when faced with competition from the United States and Canada, the two largest timber producers in the world. While NAFTA did send shocks through the community forestry sector, the worst fears were not realized, primarily because the higher quality of Mexican timber from natural forests, as opposed to plantations, continued to be demanded in the national market (see chapter 6). As of 2018, there is no documented incident of any CFE going out of business entirely because of market competition linked to NAFTA, and there is anecdotal information of communities modernizing their industrial plants in the face of the challenge. For example, Ixtlán de Juárez, in Oaxaca, saw its sales plummet 40 percent from 1994 to 1995, but it invested US$1.2 million in modernizing its sawmill (Alatorre Frenk 2000) and regained market share. In a visit I made to San Juan Nuevo Parangaricutiro in Michoacán in 2001, I found that the transition under NAFTA had been difficult because, as one of the forest industry managers told me, "our system was not very technologically advanced," but the CFE made investments to improve quality and by 1996 began to export moldings to the United States, competing successfully with Chilean producers (F. Echeverría, personal communication, January 17, 2001). In Chihuahua, the International Paper Company opened markets for small-diameter timber, creating a boom for that product (Guerrero, Reed, and Vegter 2000).

Thus, despite the indifference or hostility of the Salinas de Gortari government to community forestry, most of the reforms ended up strengthening or having no continuing negative impact on the sector. It is probable, however, that the lack of direct support slowed the sector's expansion and consolidation. A possibly more negative impact was the aforementioned article in the Agrarian Law of 1992 that allowed for the creation of the work groups (see chapter 5), weakening many CFEs, though it also served as a measure against internal corruption. But despite the clear purpose of the initiatives to try and dismantle the common property countryside, the exception given to forests allowed community forestry to survive and continue to mature in a counterhegemonic model to this neoliberal thrust (Alatorre Frenk 2000). Also in the late 1980s and early 1990s a maturing of civil society organizations supporting the community forestry sector took place and the first more academic studies were conducted. Between 1989 and 1991, Programa Pasos, a coalition of NGOs including ERA and GEA, supported by a French foundation, organized a series of workshops on community forestry (Aguilar et. al 1990; González Martínez 1992). The first publication in English on the sector occurred in 1991 (Bray 1991), and the publication of a special edition of the Mexican journal *El Cotidiano* devoted to community forestry occurred in 1992. These publications together mark the first systematic reflections by civil society actors and academics on the emerging phenomenon of CFEs. Finally, the Mexican Civil Council for Sustainable Silviculture (Consejo Civil Mexicano para la Silvicultura Sostenible; CCMSS) was founded in 1994 as a coalition of forestry NGOs and academics; it remains Mexico's most influential national forestry NGO today. Foreign assistance also began to play more of a role, with the Inter-American Foundation—where, as noted in chapter 1, I served as representative for Mexico from 1989 to 1997—supporting an array of CFEs and forest NGOs through most of the 1990s (Bray 2007). The Ford Foundation also began substantial funding of the sector in 2000 (Bray and Merino-Pérez 2002).

THE ZEDILLO PERIOD (1994–2000)

When President Ernesto Zedillo assumed office in December 1994, nearly all natural resource management agencies were, for the first time, gathered together under one cabinet roof in the Secretariat for the Environment, Natural Resources, and Fisheries (Secretaría de Medio Ambiente, Recursos Naturales y Pesca; SEMARNAP). Zedillo appointed as his environmental secretary Julia Carabias, who had been director of the National Ecology Institute (Instituto Nacional de Ecología; INE) in the last part of the Salinas de Gortari period. Carabias was an academic biologist and NGO leader with strong ties to civil society. She brought with her as part of her management team many individuals from the ranks of her own environmental NGO and other civil society activists, many of whom were active in the community forestry policy initiatives of the 1970s and 1980s. This team launched a new wave of legislative initiatives and consistent state support for community forestry that continued to varying degrees through 2018. The forestry bureaucracy itself was dropped in rank, losing its subsecretary status and becoming a directorate, but it developed several significant new forestry programs, including the Support Program for the Development of Commercial Forest Plantations (Programa de Apoyos para el Desarrollo de Plantaciones Forestales Comerciales; PRODEPLAN) and, for community forestry and small forest properties, the Forest Development Program (Programa para el Desarrollo Forestal; PRODEFOR). PRODEPLAN would continue to promote plantations, while PRODEFOR was a program of forest producer subsidies modeled on the agricultural subsidy program PRO-CAMPO, though funded at far lower levels (Merino Pérez 2004). A third program, and the second one in support of community forestry, was the Forest Conservation and Management Program (Programa de Conservación y Manejo Forestal; PROCYMAF) funded by the World Bank / Government of Mexico (Merino Pérez 2004; Merino Pérez and Segura 2005).

PRODEPLAN reflected the continued interest that the Office of the Presidency had in promoting plantations, since Zedillo's chief of staff, Luis Téllez, had been the principal overseer of the 1992 forestry law. But despite pressure from international timber interests, large plantations did not prosper, due to inadequate incentives, little clarity in the legal framework, and challenges in working with ejidos (de Ita 1996). The effort to promote plantations was given much greater resources than was community forestry, but to little effect, and the program was eventually reoriented to primarily support small plantations in communities. By contrast, community forestry had its strongest policy platform since the DGDF through the PRODEFOR and PROCYMAF programs. PRODEFOR was formally unveiled in a speech by President Zedillo in San Juan Nuevo Parangaricutiro in Michoacán on March 27, 1996, announcing the first cash grant program, direct injections of financial capital, a sign of the maturation of the sector. PROCYMAF emerged from the wreckage of a highly controversial and eventually cancelled World Bank forestry project in Chihuahua in the early 1990s and an equally unsuccessful effort by the Inter-American Development Bank (IDB) to launch a forestry program in southern Mexico (interview with S. Anta, March 20, 2017). Seeking an alternative project for Mexican forestry, the World Bank commissioned a sector study (World Bank 1995) that had recommendations focused on supporting community forestry. The new environmental secretary, Julia Carabias, adopted the study as a platform for a program in support of community forestry (A. Molnar, personal communication, June 6, 2016). PROCYMAF hence emerged as a regional pilot program with a geographic focus on Oaxaca (1997–2003) and a small-scale, flexible approach, and it later expanded to Guerrero and Michoacán (2000–2003) and Jalisco (2003) (interview with S. Anta, March 20, 2017). PROCYMAF was directly influenced both by the experiences of the DGDF in the 1970s and 1980s and by the writings of Ostrom (Segura-Warnholtz 2014). Its focus was on increasing social and human capital through training and interchanges and in diversifying forest production into NTFPs such as ecotourism, water bottling, and pine resin. It relied on four instruments of territorial planning: participatory rural evaluations, community land-use zoning (*ordenamientos territoriales comunitarios*), the legally required Forest Management Programs, and community statutes (Segura-Warnholtz 2014; see chapter 7 for more detail). It was thus an ideological successor to the earlier programs, defining its approach as "social extensionism," though it never had the number of field staff that the earlier generation of government programs deployed or their long-term presence in the communities. It also sought to better organize the forest technical service providers (Rodríguez Salazar, Anguiano Martínez, and Bray 2015). The program, under various names, continued on through the 2000s, as will be discussed further in the next section. PRODEFOR was influenced by PROCYMAF and began financing similar things nationally (interview with G. Segura-Warnholtz, May 2, 2017).

Carabias also launched a broad public consultation process for yet another forest law, since the 1992 law had come to be seen as problematic. Deregulation had gone too far, clandestine logging had increased, there were insufficient criminal penalties for violations of forest laws, and there was insufficient regulation of the forest technical services. For the plantation program, problems included failure to prohibit converting natural forests to plantations, limitations on community participation in plantation programs, and insufficient legal safeguards for the plantations (Merino Pérez 2004). The 1997 Forest Law had against this backdrop two thrusts—(1) to reregulate natural forest management and introduce more support for community forestry, and (2) to both regulate and promote new incentives for plantations—and was approved in the Chamber of Deputies in April 1997.

THE 2000S

The first two presidential periods of the 2000s, that of Vincente Fox (2000–2006) and Felipe Calderón (2006–12), represented both continuity and significant change in Mexican forest policies. At the beginning of the Fox administration, another restructuring of the forest bureaucracy took place, though support for the two major programs that subsidized community forestry, PROCYMAF and PRODEFOR, continued, and the programs were expanded significantly in the Calderón period. As well, a major new hydrological and forest environmental services program was launched in the Fox presidency and greatly expanded in the Calderón period. The Fox government restructured SEMARNAP to create the Secretary of the Environment and Natural Resources (Secretaría de Medio Ambiente y Recursos Naturales; SEMARNAT) and created new commissions for water, fisheries and forests. The National Forest Commission (Comisión Nacional Forestal; CONAFOR) was created by Fox and reported directly to the presidency. CONAFOR became responsible for forest development, though not forest regulation and approval of the management plans, which remained in SEMARNAT. This division sometimes led to inefficiencies, such as when SEMARNAT would delay approval of forest management programs that had been financially supported by CONAFOR (Merino, Ortiz, and Rodríguez 2013). Forest law enforcement was placed with the Federal Attorney General for Environmental Protection (Procuraduría Federal de Protección Ambiental; PROFEPA). The new director of CONAFOR, a politically powerful former governor of Jalisco, moved the offices to Guadalajara from Mexico City and pushed for yet another new forest law, without the wide public consultation of the 1997 law.

The General Law of Sustainable Forest Development (Ley General de Desarrollo Forestal Sustentable; LGDFS) passed in 2003, formalized the legal existence of CONAFOR (which had been created by presidential decree), brought in tighter environmental controls, and established the legal framework for an environmental services program. It also established a new apparatus of organizational control of the sector, establishing administrative units at the level of individual states called Forest Management Units (Unidades de Manejo Forestal; UMAFORs), new Regional Silvicultural Associations (Asociaciones Regionales de Silvicultores; ARS), the Program for the Strengthening of Self-Management of Forestry (Fomento a la Autogestión Silvícola; PROFAS) to support the ARS, and the National Council of Silvicultores (Consejo Nacional de Silvicultores; CONASIL), all of which frequently competed with existing regional and national organizations. This framework was an effort to create a new corporatist structure of government control of the sector but ultimately had little impact, with only a few of the UMAFORs and ARS surviving the Fox government (interview with S. Anta, March 20, 2017). However, the 2003 law did not modify the general thrust in favor of supporting community forestry. The CONAFOR budget started in 2001 with a very modest 202 million pesos, but by 2003 had jumped to 1.65 billion pesos (~US$147 million), from where it grew more slowly to 1.97 billion pesos (~US$182.4 million) by the end of the administration in 2006 (CCMSS 2008b).

The PROCYMAF World Bank program continued during the Fox and Calderón presidencies, expanding from the original three years to six, eventually investing in human and social capital in 330 communities in Oaxaca, 110 in Guerrero, and 30 each in Jalisco, Durango, and Chihuahua (CONAFOR 2012b; Rodríguez Salazar, Martínez, and Bray 2015). However, it remained a small pilot program, constituting only 3.4 percent of the CONAFOR budget by 2006. On the other hand, the budget for PRODEFOR quadrupled in the first years of the Fox administration and by 2006 constituted 17.2 percent of the budget, the second-largest percentage behind firefighting. This funding included 706 grants to subsidize FMPs, and 2,436 grants for a wide variety of training and forest development projects, including ecotourism, continuing significant investments in human and physical capital (CCMSS 2006). However, by the end of the Fox presidency, programs for production, including both community forestry and plantations, were only 27 percent of the total budget; of that 27 percent, 45 percent went to reforestation and restoration and 27 percent went to prevention and forest fires (CCMSS 2006).

The most significant new initiative that began in the Fox administration and grew even larger in the administration of Felipe Calderón, was the payment for environmental services (PES) program, which became one of the largest in the world. The program was launched in 2003 with a primary goal of preserving forest cover for water quality and quantity for downstream users, and with a secondary goal of poverty reduction. However, bureaucratic clashes and grassroots pressure diluted conservation goals, giving higher priority to poverty alleviation (McAfee and Shapiro 2010). The program responded to landowner applications (both community and individual, though most were community), and payments were contingent on maintaining forest cover. Five-year contracts were offered on a two-tier scale, with around US$36/hectare for cloud forest and US$27 for other forests (Sims et al. 2014). In the last year of the Fox administration, the program still comprised a small percentage of the CONAFOR budget, and was predominately oriented toward supporting carbon, biodiversity, and agroforestry system services (64 percent of the total ecosystem services budget), with only 36 percent for hydrological services (CCMSS 2006).

The presidency of Felipe Calderón (2006–12) gave even greater attention to CONAFOR in both programs and budget, and in both community forestry and environmental services. As well, during the period for the first time, CONAFOR began to explicitly include climate change in its strategies, noting, "the contribution of the forests to the mitigation of and adaptation to climate change" (2012a, 14). It also showed continued inspiration from the work of Ostrom, in its frequent references to the strengthening of institutions and social capital (interview with J. M. Torres Rojo, February 6, 2017). The Calderón government grouped all forest programs under the branding of ProÁrbol and more than doubled the CONAFOR budget in 2007 to 4.15 billion pesos (~US$381.1 million), and by the end of the presidential period it had a total budget of 7.182 billion pesos (~US$558.5 million; interview with J. M. Torres Rojo, February 6, 2017).

In the last three years of the Calderón period (2009–12), CONAFOR was led by Juan Manuel Torres Rojo, who had a PhD in forest economics from Oregon State University, who had been an academic researcher, and who in the 1990s had been a government forest official. Under Torres Rojo, CONAFOR developed major new efforts to support community forestry. This included new organizational initiatives such as Local Development Agencies and Centers for the Formation of Community Technicians as well as efforts to improve coordination with other government agencies. CONAFOR also funded programs

that supported women to organize microenterprises making forest-based arti-
san objects; improved silviculture; promoted certification of timber; provided
business training for CFEs; and facilitated environmental services, biodiversity
conservation, and forest carbon markets (CONAFOR 2012b). In this period,
the PROCYMAF budget took an enormous leap upward, growing nearly six
times from 2009 to 2011 to 603 million pesos (~US$43.3 million), while the
PRODEFOR budget actually shrank somewhat. However, critics noted that
the budget for reforestation and restoration was more than double than that of
natural forest management (CCMSS 2010). Also, by 2011 the PES programs
for carbon, biodiversity, and agroforestry systems were eliminated entirely for
an exclusive focus on hydrological services (CCMSS 2010). It is notable as
well that this period saw the first significant effort to rebuild the kind of forest
extension effort that was characteristic of the 1970s and 1980s. For the Formation of Community Technicians program, mentioned above, the Community
Silviculture division hired 299 forest promotors, normally from the one of the
communities or from the region, to provide technical and organizational support for projects and to carry out community assessments (*diagnósticos*) (Anta
Fonseca 2015), though this activity was still less than the intensive extension
work characteristic of the 1970s and 1980s. At the same time, PRODEFOR
and later PROCYMAF continued the practice of endowing communities with
physical capital, buying sawmills and other logging equipment. CONAFOR
also carried on the plantation program launched during the Salinas de Gortari period, though with a reorientation toward small community plantations,
establishing 6,925 plantations on over 375,000 hectares, an average of fifty-four
hectares (CONAFOR 2012a).

Due to PROCYMAF's dramatic growth in the last years of Calderón's
presidency, the program merits special attention. As noted earlier, the program
bridged the Fox and Calderón administrations, with the so-called PROCY-
MAF II program running from 2003 to 2008, still with World Bank funding.
The program then became the Community Forest Development Program and
finally Community Silviculture with exclusively Mexican government money
by 2010 and with a national presence. The program handed out grants to 2,256
ejidos, comunidades, and intercommunity organizations, covering 4.6 million
hectares (CONAFOR 2012b). The foci of the PROCYMAF programs and
its successors echoed those of the DGDF and other government programs
from the 1950s to the 1970s, though without the major extension presence. It
developed ambitious programs, with a total of thirteen major lines and with a

significant focus on human, social, and physical capital. Two of the project lines focused explicitly on developing community rules around land use, including the land use zoning and community regulations program (Anta Fonseca et al. 2006), and six were focused on training and human and social capital formation (Torres Rojo and Amador Callejas 2015b). Salvador Anta Fonseca (2015), who was the director of the Community Forest Development Program and Community Silviculture from 2010 to 2013, has documented CONAFOR's investments in social and human capital. From 2006 to 2012 it supported 924 ordenamientos territoriales comunitarios covering 6.3 million hectares, 779 participatory rural evaluations, and 700 workshops to develop community regulations or rules, again programs directly inspired by Ostrom's writing. Studies to establish community-conserved areas (see chapter 7) were conducted in 179 communities covering some 500,000 hectares, while grants for Forest Stewardship Council (FSC) certification covered 879,043 hectares, and new forest associations were promoted (Anta Fonseca 2015).

Despite this major new support to community forestry, the bulk of the CONAFOR budget, as the CCMSS has noted, went to reforestation, restoration, and payment for environmental services, not community forests. Figure 6 shows the major CONAFOR budget categories for 2003–11. The dotted line represents reforestation and restoration, and beginning in 2006 with the Calderón presidency it soared to become far and away the principal expenditure, since reforestation was one of Calderón's signature programs. Payment for environmental services (solid line) emerged for the first time in 2004 and by 2010 was the second largest budget item. In 2008–9, commercial plantations were the second most important program. The category natural forest management, which includes the PRODEFOR and PROCYMAF programs (dark, tightly dashed line), got increased funding in 2006 and, as noted above, took a sharp leap upward in 2011, a tendency that continued in 2012. Thus, in the last two years of Calderón, PROCYMAF and PRODEFOR came to occupy second place, in funding, or nearly 30 percent of the budget, bringing a new focus on CFEs in some of the most important community forestry states (Zúñiga and Descamps 2013). Far below reforestation, they nonetheless constituted major new funding for the sector.

In addition to the very substantial expansion of state support to community forestry in the Calderón period, the government during that time was also notable for the launching of Mexico's REDD+ program. REDD+, introduced in chapter 1, is the global strategy of reducing carbon emissions derived from

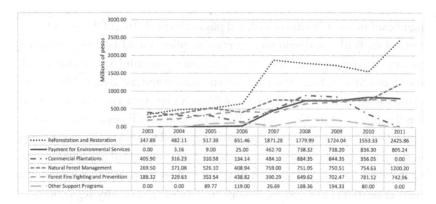

	2003	2004	2005	2006	2007	2008	2009	2010	2011
······ Reforestation and Restoration	347.89	482.11	517.39	651.46	1871.28	1779.99	1724.04	1553.33	2425.86
—— Payment for Environmental Services	0.00	3.16	9.00	25.00	462.70	738.32	738.20	836.30	805.24
— · · Commercial Plantations	405.90	316.23	310.58	134.14	484.10	884.35	844.35	356.05	0.00
— — Natural Forest Management	269.50	371.08	526.10	408.94	759.00	751.05	750.51	754.63	1200.20
— — Forest Fire Fighting and Prevention	188.32	229.63	353.54	438.82	390.29	649.62	702.47	781.12	742.96
——— · Other Support Programs	0.00	0.00	89.77	119.00	26.69	188.36	194.33	80.00	0.00

FIGURE 6. CONAFOR budget categories (2003–11). Graph adapted from Zúñiga and Descamps (2013).

conversion of forests to other uses. The program is designed to be a compensation mechanism from rich countries who have excess carbon emission production to poor countries, rich in natural resources and with lower emissions (Angelsen 2009). Mexican forests briefly occupied the global spotlight in 2010, when Mexico hosted the Sixteenth Annual Conference of Parties (COP) of the Kyoto Protocol. I will further discuss the REDD+ program as part of climate change and the future of Mexican forestry in the concluding chapter.

The government of Enrique Peña Nieto (2012–18) marked the first time since 1994 that forest policy in support of community forestry had been relatively less prominent, focusing more on increasing production. The Peña Nieto government's National Strategy for the forest sector included, among other things, an 86 percent increase in timber production, from 5.9 million cubic meters to 11 million cubic meters, but also called for a tripling of certified forests to 2.5 million hectares, the establishment of 750 new CFEs, and the incorporation or reincorporation of 4.6 million hectares into management to increase production, with criteria of biodiversity conservation and forest zoning (Benet 2018; Gobierno de la República 2014). His government also continued the program of forest extension established in the previous presidential period, creating 750 community promoters to work in a similar number of forest communities (E. A. Ellis et al. 2015). Forest community members from throughout Mexico were brought to "Instructor Communities," such as San Pedro el Alto in Oaxaca, for advanced training, a noteworthy effort to rebuild a forest extension capacity. However, an analysis of the 2015 budget for CONAFOR under Peña Nieto found it to be "opaque" and that it represented "a clear backwards step in transparency and accountability with respect to the years before 2012" (CCMSS 2014, 5). As well, a drop in oil prices led in 2017 to a 40 percent cut to the CONAFOR budget. The government fell short on all of the announced goals. By 2017 it was estimated that forest production increased modestly to 7.6 million cubic meters (far short of the 11 million goal), forest certification increased little, only 259 new CFEs were created (34 percent of the goal, including both timber and NTFPs), and only 55 percent of the stated goal was achieved in increased production areas (Benet 2018; CCMSS 2016; Deschamps Ramírez and Madrid Zubirán 2018). In addition, the Community Silviculture program (the successor to PROCYMAF) was eliminated entirely. Support to production and community forestry, reduced to around 15 percent of the budget, consisted of most funds going to "passive conservation" of the environmental services program and reforestation; subsidies were focused on particular states and municipalities without a clear strategy (Deschamps Ramírez and Madrid Zubirán 2018). Most worrisome is the evidence that deforestation, which had been in decline since the late 1990s, jumped dramatically under Peña Nieto from 158,000 hectares annually to 299,000 hectares annually, principally in the tropical forests of Chiapas and Campeche (Benet 2018; see chapter 7 for more on deforestation).

One notable event during the Peña Nieto period was a new forest law, passed in the Chamber of Deputies in April 2018. The law had a torturous legislative process, and was not considered a substantial advance over the 2003 law. However it did establish, for the first time, that *manejo forestal comunitario* (community forest management) was an official policy of the government with a criteria of "affirmative action" on its behalf, added additional language on sustainability and an ecosystem focus, attempted to reduce regulation, and instituted new measures of control over timber retailers and a requirement that all government timber purchases have transparent supply chains to combat illegal logging (Benet 2018; G. Chapela 2018). In the new government of Andrés Manuel López Obrador, who assumed power on December 1, 2018, León Jorge Castaños Martínez, the most important reformer of the 1970s and 1980s, was back as director general of CONAFOR. Although this change suggested a renewed emphasis on community forest management, significant budget reductions in an austerity move and the transfer of the reforestation program to a new government welfare agency also suggested a CONAFOR diminished in size and influence.

As noted at the end of chapter 2, the community forest sector largely came into being by 1988, with some 1,300 communities producing timber, and by

2011–13 there had only been a relatively modest increase to 1,621, with some flux in CFE start-ups and the demise of existing ones. With the exception of the Salinas de Gortari period, state policy during this period has been characterized by continued and generally increasing support until recent years. Common property theory has not taken into account such sustained state policy support in the supply of institutions and the five capitals necessary for common property enterprises to emerge and consolidate and to form the economically distinctive and resilient SES of today. These successful reforms from above were complemented by grassroots mobilizations to claim and expand economic spaces in some states, a vigorous entrepreneurial response to the openings by many hundreds of other communities, and, accompanying the disappearance of the parastatals, full engagement with market prices (to be more extensively reviewed in chapter 6)—all important and historic achievements. Nonetheless, analysts have also found significant weaknesses in the sector. For example, as we saw earlier in the chapter, since CONAFOR was created in 2001, the budget increased by more than 3,000 percent (see figure 6), but national timber production has gone down by around 20 percent, the share of forestry in the GDP was reduced to 0.3 percent by 2011, and the trade deficit in timber increased by some 133 percent from 1997 to 2006 (García López 2013). In the remainder of this chapter, I will review these issues and others with respect to the performance of the community forestry SES in the late 2010s.

THE STATUS OF THE SES OF MEXICAN FORESTS AND COMMUNITY FORESTRY AS OF THE LATE 2010S

To understand the complexity and state of resilience of Mexico's community-managed forests as of the late 2010s, it is useful to explore a series of questions: (1) What is the magnitude of the SES? That is, how much of Mexico's forests are owned by communities, and how much is under management by how many communities? (2) What is the significance of the community forestry SES to the Mexican economy and the balance of trade? (3) Has the SES played a role in reducing the major problem of illegal logging? (4) It is frequently asserted that the growth of the SES is inhibited by overregulation. What is the state of government regulation of Mexico's forests, and what is regulation's impact on the SES? (5) What is the status of NTFP production and small forest properties as important components of the forest sector? Issues of the creation, growth, and mortality of CFEs will be covered in chapter 6, on economics; the impact on deforestation will be covered in chapter 7; and the future prospects for expansion of the sector will be covered in the concluding chapter, 8.

OWNERSHIP OF MEXICO'S FORESTS BY COMMUNITIES, NUMBER OF FOREST COMMUNITIES, FOREST MANAGED BY COMMUNITIES

The current estimate of tropical and temperate forest cover is 65.3 million hectares, or almost exactly one-third (33.57 percent) of the national territory, nearly evenly divided between temperate forest (51.7 percent) and tropical forest (48.3 percent). In the 1950s Hinojosa Ortiz (1958) estimated that 18 percent of the forests of Mexico were on community lands. However, beginning around 1980, the commonly cited figure took a great leap upward to 80 percent. However, there was no empirical basis for this estimate, despite its widespread use. A series of careful studies have recently converged on a figure of around 60 percent of the forests in Mexico owned by communities. Madrid et al. (2009) found that 60.3 percent of the forests are on community lands, with Morelos, Oaxaca, and Guerrero all having over 80 percent community ownership. M. Skutsch et al. (2013) proposed a figure of 58.8 percent for forests on community lands, and Torres Rojo and Amador Callejas (2015a) found 61.8 percent of forests on community lands for 2007, with nearly equal amounts of temperate and tropical forests and an average community forest cover of over 60 percent. These three studies allow for considerable confidence in saying that communities own around 60 percent of Mexico's forests. If the 1950s figure is correct, the forest distributions of the Díaz Ordaz and Echeverría presidential periods (1964–76), which resulted in major waves of titling of forests to communities in Durango and Chihuahua, merit much more credit than they have received for the forest ownership that is the foundation of Mexican community forestry. In this period, the agrarian reform led to a huge expansion of forests on community lands, from 18 percent to probably close to the 60 percent of today, an enormous endowment of natural capital to Mexican forest communities.

The next question is how many "forest communities" own the 60 percent of forests? The question should be divided into how many communities have at least 5 hectares of forest, and how many of those communities are managing their forests for timber (leaving aside the number who manage them exclusively for NTFPs). As of 2012 and as noted in chapter 1, there were 31,837 agrarian communities in Mexico (29,490 ejidos and 2,347 comunidades) occupying

100,308,240 hectares (51 percent of the national territory). Private property followed, holding around 70 million hectares (35.7 percent), and government lands (federal, state, and municipal) held 11.5 percent (Robles Berlanga 2012). J. M. Torres Rojo and J. Amador Callejas (2015a) analyzed all ejidos and communities that had more than 5 hectares of forest cover and any type of vegetation, though there were only a small number below 100 hectares. Their figures suggest large numbers of agrarian communities with significant amounts of forest. Table 2 below reworks some of their figures, suggesting that there are 4,580 agrarian communities with an average of 2,037 hectares of temperate forest, 7,680 with an average of 1,543 hectares of tropical forest, and smaller numbers with mixed forests (temperate, tropical, and/or arid/semiarid) on their lands.

These data suggest a total of 17,586 agrarian communities with at least 5 hectares of forest and a total area of over 49 million hectares, each community averaging 2,793 hectares. However, many of these communities appear to have mostly noncommercial forests, such as oak forests or dry tropical forest. Some 46 percent of the estimated 20.3 million hectares of community temperate forests are oak or montane tropical forests, and 52 percent of the 18.2 million hectares of tropical forest are dry tropical forests (Torres Rojo and Amador 2015a, 26, calculated from table 1.5). It is also useful to note that distribution of forests among all communities with forests is highly unequal. For example, in all types of temperate forests, only 8.2 percent of the communities have more than 5,000 hectares, and these have 69.4 percent of all forests. In tropical forests,

TABLE 2. EJIDOS AND COMMUNITIES BY FOREST TYPE, TOTAL FOREST, AND AVERAGE SIZE OF FOREST

FOREST TYPE	TOTAL EJIDOS AND COMUNIDADES	TOTAL FOREST (HECTARES)	AVERAGE SIZE OF FOREST (HECTARES)
Temperate	4,580	9,329,942	2,037
Tropical	7,680	11,850,429	1,543
Temperate and tropical	3,173	14,895,005	4,694
Temperate and Arid/semiarid	1,246	6,018,476	4,830
Tropical and Arid/semiarid	721	5,331,985	7,395
Temperate, Tropical, and semiarid	186	1,695,693	9,117
Total	17,586	49,121,530	2,973

Source: Adapted from (Torres Rojo and Amador Callejas 2015a, 27, table 1.6).

10.2 percent have more than 5,000 hectares, and these have 60.6 percent of the forests, reflecting the luck of the agrarian draw (Torres Rojo and Amador Callejas 2015a, calculated from table 1.6). Available evidence also suggests a strong correlation between the vertical integration typology and the size of forest under management. As noted in chapter 1, Bray et al. (2007) found in a sample of 879 communities in the ten most important forest states that Type IIs had an average of 922 hectares, Type IIIs 1,553 hectares, and Type IV 3,503 hectares under management in their current management plans.

So how much of all these forests on community lands are commercially viable? CONAFOR (2012b) estimates that there are some 15 million hectares of commercial forests in the country, or around 30 percent of total tropical and temperate forests. Of this area, CFEs have about 6.2 million hectares in Mexico under management (Torres-Rojo, Moreno-Sánchez, and Mendoza-Briseño 2016), suggesting that only around 40 percent of commercially viable forests are being logged, with substantial implications for forest conservation explored in chapter 7. However, the basis for the estimate of total commercial forests or where they are located is not clear. For example, O. Estrada (2018) has noted that for Chihuahua, the second most important forest production state, nearly all of the viable commercial forests are being logged.

In any event, how many CFEs are managing the estimated 6.2 million hectares? There is a definitional issue as to what constitutes a CFE that has led to some of the variation in the reported numbers. It has been suggested, for example, that Type II stumpage communities are not operating CFEs, since some number do not do anything but take the money for selling the timber off their land, keep no accounting, and have no capitalization. In this interpretation, a CFE only emerges when the business is formally incorporated, when a manager or managerial council is established, and when other aspects of a formal business operation are achieved. For example, F. Cubbage et al. (2015), citing a CONAFOR source, suggests there are only 992 CFEs, but CONAFOR is not counting Type IIs as enterprises. The definition used in this book is that all communities that produce timber are CFEs. Some may be enterprises in which no operating capital is maintained, in which all profits are immediately distributed, and which shut down entirely between the logging seasons, but they are nonetheless enterprises based on a common property for which they carry out productive activities and realize income in the marketplace. Based on this more encompassing definition, a study of government logging permits from 1992 to 2002 found that precisely 2,300 ejidos and comunidades had received at least one logging permit over the entire period, including 1,867 ejidos and 433 comunidades (Bray et al. 2007). The number of communities with permits in any given year during this period will be smaller, since not all communities with authorizations log every year, for a variety of reasons. As mentioned in chapter 1, a snapshot of the typology for 2011–13 by Torres Rojo and Amador Callejas (2015a) found that there were 699 Type II (43 percent), 738 Type III (46 percent), and 184 Type IV (11 percent), for a total of 1,621 communities with authorizations in that period. This would appear to be the best current estimate of how many CFEs are operating in a given year, though Carillo-Anzures et al. (2017) arrived at higher numbers. Communities are normally classified at the highest level in which they participate, but it should be noted that these types conceal the fact that many communities sell product mixes of stumpage, roundwood, and sawnwood. In the Michoacán/Durango sample, several communities sell both stumpage and sawnwood or sell both roundwood and sawnwood. As well, some communities may sell roundwood one year and sawnwood the next, depending on community economic strategies (Antinori and Rausser 2010). As will be discussed more in chapter 6, considerable flux of active CFEs happens from year to year, but this fluctuation also suggests relatively modest overall growth of the sector after the late 1980s.

FOREST PRODUCTION IN THE MEXICAN ECONOMY

A major reason that the political economy of forest reforms have been so successful in Mexico is that it doesn't challenge important economic interests. In Indonesia the forest sector accounts for 3.5 percent of the GDP (ITS Global 2011), while in Mexico it is far less. From 2003 to 2012 the forest sector's share of the Gross National Product (GNP) was an average of 0.17 percent, and forestry's participation in the agriculture, livestock, and forest sector an average of 4.61 percent (Gobierno de la República 2014). Historically, only around 4 percent of the total budget for agriculture was allocated to the forestry agency (World Bank 1995). As well, roundwood production, the production of logs, of which 80–90 percent is from communities, has been highly uneven since 1987, though with downward tendencies since 2000 (see figure 7). Going back earlier than what is shown in the figure, in the early 1950s annual timber production averaged 3.68 million cubic meters but had a strong upward growth curve in the 1970s and 1980s, reaching as high as 9.37 million cubic meters by 1987 (Caballero Deloya 2008, 2010). As we have seen, this is the period when the CFE sector was created and matured, and it likely that this growth was a contributing factor. The rise has also been attributed to exchange rate issues and economic crises in the 1970s and 1980s, increasing the costs of imported timber (personal communication, J. M. Torres Rojo). Since the high point at the end of the de la Madrid administration, however, production has been highly irregular but with downward tendencies.

The significant drop in production during the presidency of Salinas de Gortari (1988–94) can be attributed to both an almost complete policy neglect of the community forest sector and the shift to open markets, increasing imports of timber. The period 1995–2000, when production took a sharp rise upward, can be attributed both to the PROCYMAF and PRODEFOR programs but also a boom in demand for packaging for fruit and vegetable exports. However, the equally sharp declines in production in the Fox administration (2000–2006) is linked to a reduction in authorizations during the establishment of CONAFOR and the reorganization of SEMARNAT in the first two years of the that presidency, leading to a rise in imports of sawnwood. The slight recovery in 2007 to some 7 million cubic meters may be due to state support, but the renewed decline after that has been attributed to the strong role of illegal timber in agricultural commodity export packaging, as well as the increased use of plastics in packaging. The 2003 Forest Law had no discernible positive impact on production, since production had already dropped by 2001, and production

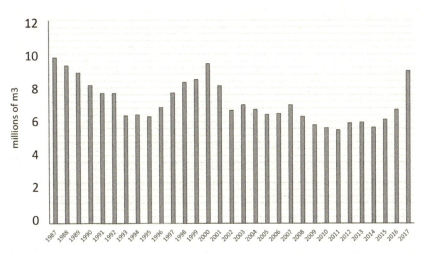

FIGURE 7. Roundwood production in Mexico (1987–2017). Graph from Gobierno de México n.d.

continued to trend downward, reaching a twenty-year low of 5.5 million cubic meters in 2011, a 41 percent decline from 2000. Both the PROCYMAF (under various names) and PRODEFOR programs continued and expanded in the 2000s but were not particularly focused on increasing production (J. M. Torres Rojo, personal communication). The 2017 rise in production to an estimated 7.6 million cubic meters (Deschamps Ramírez and Madrid Zubirán 2018) is not an official figure but if correct suggests that the efforts by the Peña Nieto government to increase production had some effect.

The general tendency toward declines in production from 2000 to 2014 varied by state. For the 2000–2010 period, production in Durango, the most important forestry state, fell by 29 percent, below the national average, while production in Chihuahua fell by 41 percent, right at the national average. However, production in Jalisco fell by 55 percent and in Michoacán by 68 percent, in the latter state possibly as a result of the expansion of avocado groves. In Oaxaca, production fell by only 14 percent and in Quintana Roo by 16 percent, and production actually rose in Puebla (4 percent), Veracruz (13 percent), and Tamaulipas (30 percent) (Gobierno de México n.d.). The number of logging authorizations also dropped by over a third from 2002–11 even as production was relatively stable at a new lower level, suggesting larger volumes produced per management unit. This shift is likely due to a decline in authorizations for small private producers, suggesting a crisis in that sector that may have also driven a rise in illegality, to be discussed in the next section (J. M. Torres Rojo, personal communication). Of particular note is that harvests are far below the annual authorized volumes, around half of the nearly 12 million cubic meters authorized per year (CONAFOR 2014), suggesting both inefficiencies and conscious efforts to conserve on the part of communities. In addition to the decline in production, harvest rates are also low, at 2.5–4.2 cubic meters / hectare / per year when the potential productivity of the forests of Durango are estimated at 20–25 cubic meters / hectare / per year (Corral-Rivas et al. 2016).

Mexico also runs large and persistent trade deficits in forest products. Harvest volumes in the late 2010s constitute some 17 percent of forest product consumption, with imports in pulp, paper, plywood, and fiberboard composing most of the trade deficit of close to US$6.2 billion in 2016 (Torres Rojo n.d.). However, Mexican sawnwood is more competitive; it has about 50 percent of market share, due to its higher quality over imported plantation timber. The national and international competitive position of Mexican forest products will receive further review in chapter 6. There is evidence that the persistent and large deficits in timber products are a matter of policy; a 2006 official report suggested that business opportunities for Mexico in the forest sector are more in the transformation of imported raw forest materials "extracting the natural resources of others and conserving our own" (SEMARNAT/CONAFOR 2006, 11). However, Torres Rojo (n.d.) has argued that the declines in production are significantly due to competition from illegal logging, which takes us to the next section.

IMPACTS OF ILLEGAL LOGGING ON CFES AND SOURCES OF ILLEGAL LOGGING

Illegal logging is considered one of the most difficult issues facing the Mexican forest sector (G. Chapela y Mendoza 2018a). There have been a wide variety of estimates of illegal logging and the geographical regions in which it may be concentrated. A report by Greenpeace estimated the illegal production at 7 million cubic meters, roughly equivalent to the legal production (CCMSS 2007), while a more recent estimate put it at 70 percent (G. Chapela y Mendoza 2018a). Miguel Caballero Deloya (2010) estimates that illegal logging is much higher than is normally estimated, although he includes unregulated firewood extraction. Official government figures from 2009 suggest that illegal logging represents only 30 percent of the authorized production and is responsible for 8 percent of total deforestation (Gobierno de la República 2014). The most reliable recent estimate, using novel methods based on trade figures, proposes that the average magnitude of illegal logging for the 1999–2014 period is around 5.7 million cubic meters, similar to the average annual harvest in that period (Torres Rojo n.d.). Illegal logging appears to be concentrated in particular states. PROFEPA has identified 108 "critical deforestation zones" for illegal logging scattered throughout the country but with the highest number in Veracruz, Jalisco, Querétaro, Mexico State, Puebla, Chiapas, Guerrero, and Mexico City (Buendía 2016). For the most part, these are not the major forest production states and the majority of the zones are reported to be in protected areas, suggesting that PROFEPA is not monitoring illegal logging hotspots outside of protected areas (Enciso 2016). Illegal logging in the Monarch Butterfly Biosphere Reserve has received much attention, though it has diminished substantially from earlier periods (Vidal and Rendón Salinas 2014).

It appears that the bulk of illegal logging comes from communities without forest management plans, from communities with management plans that have parceled their forests, from protected areas, and from small forest private properties. In the majority of CFEs with authorized FMPs, illegal logging is reduced due to community control (J. M. Torres Rojo, personal communication). In a sample of forty-four forest management communities in Oaxaca, it was found that problems with illegal logging were concentrated in five communities that had parceled their forests individually, indicating weak community forest governance (Antinori and Rausser 2007). However, a study in Michoacán and Durango reported that illegal logging occurred in 29 percent of the Type II communities and 17 percent of the Type IV communities (Antinori and Rausser 2010), suggesting that forest management communities are not exempt from the problem. A significant overcapacity in processing infrastructure may also contribute to illegal logging. There are an estimated 9,000 timber-processing facilities with capacity to process three times the annual authorized volume. As well, PROFEPA is more vigorous in enforcement of communities that have legal authorization than it is of these processing centers. From 2001 to 2010 only 10 percent or less of 10,163 inspections that PROFEPA reported were in sawmills or other timber-processing centers. Despite these gaps, the amount of illegal timber confiscated declined sharply over the period (CCMSS 2012). Thus, a wide variety of causes have been ascribed to illegal logging, including overregulation, poor community governance, economic inequities within communities, poor law enforcement, declaration of protected areas over community land and, for small private properties, a high degree of absentee ownership (CCMSS 2007). Torres Rojo (n.d.) argues that the high volume of illegal sawnwood is the principle reason for the lack of growth in national legal production. However, the most disturbing cause of illegal logging is the increasing involvement of organized crime in the sector. That subject and the consequent violence will be further discussed in chapter 5.

ARE MEXICAN FORESTS OVERREGULATED?

Overregulation is commonly asserted as the most important reason for declines in production and other problems in the community forest sector (Deschamps Ramírez and Madrid Zubirán 2018). Mexico indeed has a strong legal regulatory framework. However, the period of deregulation, from the 1992 Forest Law to the 1997 Forest Law, is widely seen to have led to a rise in illegal logging. The 1992 Forest Law deregulated timber transportation by ending the requirement for written documentation, and the deregulation also apparently led to misunderstandings in communities. In a study of a two communities in Michoacán, it was found that they misinterpreted the 1992 amendment to Article 27 to mean that they could proceed to parcel their forests and manage them individually. This misunderstanding, combined with the deregulation of transportation, initiated a surge in illegal logging in the communities. It was estimated that between 1992 and 1995, up to forty logging trucks hauled timber out of the two study communities daily, wiping out most of the forests, with the timber being sold at extremely low prices (Barsimantov and Navia Antezana 2012). However, since the reregulation of transportation in the 1997 Forest Law, it is a common observation that the sector is overregulated. To evaluate the issue of overregulation, I will first review the Mexican forest regulatory apparatus. As of the mid-2010s there are three levels of regulation of commercial forest extraction, for both timber and non-timber forest products (Bray and Duran-Medina 2014):

- LGDFS, modified in 2008 and 2013.
- The Reglamento (regulations) of the LGDFS, composed of 178 articles in forty-four pages, from 2005.
- Number 152 of the Norma Oficial Mexicana (Mexican Official Norm; NOM) (SEMARNAT/CONAFOR 2006), which establishes the guidelines, criteria, and specifications of the FMPs for timber and non-timber forest harvests. The NOM-152 was approved in 2008 and is composed of twenty-four pages of detailed instructions. This is the document that provides the framework for the elaboration of the legally required FMPs.

The CCMSS has argued that "normativity and bureaucracy have combined to undermine the potential of the forest sector, and impedes its contribution to the search for socially inclusive, economically solid and environmentally responsible growth for Mexico. In a few words, overregulation has become an obstacle to the sustainable development of the country" (Fernández Vázquez and Mendoza Fuente 2015, 4). A forest community in Jalisco supported this argument by noting, "they make us jump through hoop after hoop . . . [M]any communities now have gotten dispirited and have stopped trying to log their forests by legal means" (CCMSS 2008b, 4). In Quintana Roo excessive requirements for forest management from SEMARNAT are a common complaint, exacerbated by the requirement for separate documentation of environmental impacts for tropical forests (E. A. Ellis et al. 2015). However, the forest production statistics above do not suggest that legislation has any link to declines in production. The 1997 law actually reregulated forestry production, and production rose after that; furthermore, the 2003 law added environmental regulations, with little discernible impact on production in the immediate years after.

The most thorough study of the regulatory process was commissioned by the CCMSS (García Aguirre 2014). This study rigorously documents a very large number of bureaucratic steps that are required for logging approval in Mexico, divided into the "Prior Process" and the "Environmental Process." The Prior Process includes Assembly approval and copies of titles, copies of community statutes, internal written rules, and the development of the FMP. In all, the Prior Process involves seventeen to twenty-one separate steps, with the additional steps being required for Environmental Impact Statements in tropical forests. No time limits are specified for this process, but once the FMP is completed and submitted, the Environmental Process, the formal approval, begins. This phase involves more than twenty additional steps, all of which are supposed to take place in a short period of time. Not surprisingly, the number of steps and the difficulty in completing them rapidly can cause delays in the approval process. García Aguirre notes frequent complaints about long bureaucratic delays but, in his words, "the exact frequency or average length of these delays has not been documented and the seriousness of these delays varies by state" (García Aguirre 2014, 51).

The process is also exacerbated by a shortage of personnel to carry out the steps, particularly in states such as Michoacán that have large numbers of small private forests (García Aguirre 2014), so the problem may be understaffing as much as overregulation. As further burdens on staffing, there are additional paperwork requirements for storage centers, transportation, phytosanitary regulations, sales, and annual reports at the end of each harvest season, all of which imply costs for the producers. The direct costs are relatively minor, but the indirect costs for community authorities in time and travel to follow up on delayed authorizations can raise expenses to as high as nearly 25 percent of the value of the timber. Nonetheless, most of these costs can be borne by ongoing operations, and complaints about overregulation are not universal. For example, in an interview I conducted in Durango in February 2019, a community authority, when asked about regulations, said, "They're not so bad, and they help us to maintain and regulate the life of the forest. In other areas there is a lot of clandestine logging."

Further, forest regulation in Mexico also takes place in a context of, as we have seen, the substantial devolution or decentralization of rights over the forest resource. R. F. Hajjar, R. A. Kozak, and J. L. Innes (2012) conducted a comparative assessment of the amount of perceived power that communities have over managing forests in Mexico and Brazil. In a study of three communities in tropical forests in the Yucatán Peninsula, they found that community members in Mexico were more satisfied with their degree of management authority than in the Brazilian case studies. Despite heavy regulation of the extraction process, communities have learned to follow the complex rules, and they have the relative power to cut below the annual volume and over what species to log, as well as substantial power over subsistence uses and rules on benefits (R. F. Hajjar, R. A. Kozak, and J. L. Innes et al. 2012). The regulations may indeed represent "a very important entrance barrier" to CFE startups (García Aguirre 2014, 51). An overlooked aspect of overregulation is the focus on forest inventories and not the postharvest condition of the forest. It has been suggested that this shift in perspective could make much paperwork unnecessary (interview with J. M. Torres Rojo, February 17, 2017). In summary, the calls for deregulation do not seem to take into account the experience of deregulation in the 1990s and overlook the long list of other factors that appear to be inhibiting the further expansion of the sector. Mexico CFEs are heavily and likely overregulated and are one area in which the political economy of reform has not been felt, but overregulation is not necessarily a major problem for existing CFEs and may be most relevant in the launching of new CFEs (see chapter 8). Nonetheless, regulation is almost certainly heavier than it needs to be, and a simplification of the process—particularly for smaller forests, small private forests, and start-up CFEs—is necessary.

COMMUNITY ENTERPRISES FOR NTFPS AND SMALL PRIVATE FOREST PROPERTIES

The focus of this book is on commercial timber production from community forests, but some communities are increasingly organized around the production of NTFPs and merit a brief discussion, along with small private forest properties. A study of NTFP authorizations in twelve states from 2002 to 2009 found that 810 permits for NTFP extraction were given to agrarian communities and another 468 to small private forest properties. Products included pine resin, soil, mushrooms, moss, palm buds, and barbasco rhizomes, among many others (Carillo-Anzures et al. 2017), with bottled spring water an important new NTFP in some states. Chicle in Quintana Roo is a well-known traditional

NTFP that faced declining markets after the post–World War II invention of petroleum-based chewing gum but has recently found modest new markets as a traditional sustainable forest-based product (Chizcagum 2018; Konrad 1987, 1991; Redclift 2004). Many traditional NTFPs are extracted by individuals from community lands by permission of the Assembly, the only collective action occurring in marketing cooperatives and allotment of extraction areas. However, bottled spring water and the nonextractive service industry of community-based ecotourism, promoted by PROCYMAF beginning in the 1990s, are strongly emerging community enterprises based on the common property, and they occur as both a diversification from timber and as a new form of community forest enterprise. As well, the emergence of voluntary conservation areas (see chapter 7) suggests that there are now large numbers of communities who conduct some degree of forest management exclusively for strict protection. The increasing numbers of communities who are not extracting timber but engaging in NTFP production, ecotourism, or conservation argue for the need for an additional CFE category (Type VI) for communities who engage in these forms of forest management (Hernández-Aguilar et al. 2017), as noted in chapter 1 and to be discussed further in chapter 5.

A final highly relevant subject is that of the small private forest properties. Mexico has a sector of what in the United States would be called nonindustrial private forests (NIFPs) or "family forests." As noted earlier in the chapter, these are referred to legally as pequeñas propiedades forestales, with an upper limit of 800 hectares. There is little literature on them, and they are the most significant unexplored area in Mexican forest management. However, F. C. Carillo-Anzures et al. (2017) have recently shed new light on the magnitude of the sector. In a study of forest producers in the twelve most important forest states in Mexico, the authors found that there were 7,556 small forest properties in total. The three most important states were Veracruz (1,660 properties), Michoacán (1,447 properties), and Puebla (1,035 properties). By the researchers' estimate, of the total forest area of ejidos, comunidades, and small forest properties under FMPs of 9.8 million hectares, nearly 10 percent was in the small forest properties but with 22.4 percent of the authorized volume, suggesting these properties had much higher productivity than that of the community forests. However, it is not known how much of this authorized volume was actually harvested. One of the earliest documented efforts at improved forest management emerged from a group of small forest properties in Tlaxcala led by the private forest owner Carlos Caballero, that served as an inspiration for the DGDF (Castaños Martínez

2015). The organization pioneered a silvicultural practice of small patch clear-cuts and in 1974 formed the Association of Silviculturists of the Municipality of Tlaxco (Asociación de Silvicultores del Municipio de Tlaxco; ASILVIT-LAX). It endures as of 2019, comprising thirteen small ejidos and nineteen small private properties collectively owning 29,689 hectares with management plans covering some 5,200 hectares of forested land (B. Hodgdon, personal communication, March 12, 2019). There have also been multiple state initiatives at placing agrarian communities and small private forests in the same organization, such as the Uniones de Permisionarios (Permit-holders Unions) in Durango, to facilitate the provision of forest technical services and other support.

CONCLUSIONS

The period 1988–2018 represented the consolidation and maturation of the CFE sector. By 1988 the major victories for forest tenure had been won, and by 1994 Mexican forest policy shifted into decided and ongoing support for community forestry, with little need for continued community mobilizations demanding it. The period of Salinas Gortari constitutes the only presidential period since 1970 when there was no significant support in programs for the sector, and even during this period the constitutional reforms gave communities new ownership rights over the forests and greater autonomy in community management in general. Beginning in 1994, new support programs deepened and expanded the degree to which Mexico was a facilitative political regime and solidified the transformative decentralization of the sector. The reforms and programs from above created the defensive perimeter within which self-governance of the community forest enterprises could develop and community collective action could take place, mitigating the impact of neoliberal policies that devastated smallholder agriculture (Fox and Haight 2010). State support was always less than many in the sector desired or thought appropriate, but nonetheless is well beyond what occurred in any other developing country. Communities own some 60 percent of the forests of Mexico, some 6.2 million hectares under management, and millions more under various forms of formal and informal conservation (see chapter 7). Important factors that have allowed the sector to emerge were the persistent ideals of the Mexican Revolution embodied in top-down reformers and the fact that forestry is not important to the Mexican economy. Illegal logging is likely a significant depressing factor on legal production, and further expansion of the community forest sector could help control it. Finally, a new and vigorous sector of community forests based on NTFPs and conservation has become a new presence on the scene, and private forest properties are an important but understudied sector of Mexican forestry. In the next chapter, I will flesh out how state policy, markets, and vigorous community collective action play out in two extensive case studies and issues related to institutions and organizations.

CHAPTER 4

THE LARGE SHADOW OF COMMUNITY

Varieties of Experiences in Mexican CFEs

Today, many successful Mexican CFEs display relatively high levels of coordination and collective action around the administration of industrial forestry. Communities displayed astonishing entrepreneurial vigor in occupying spaces created by the state, in demanding new spaces where they did not exist, in influencing forest policy, and in taking full advantage of the many subsidies granted by the state. In this chapter and the following one, I will examine more closely the vigorous response of strong community governance and the self-governance of CFEs in response to market incentives and state policy. Successful CFEs have highly heterogeneous origins, from communities with millennial traditions of territorially based indigenous communalism to populations of random groups of colonists or family networks who quickly formed a bond with a territory. In almost all cases these were communities of subsistence corn farmers, whose traditional forest use was limited to gathering of firewood, obtaining timber for domestic use, using medicinal plants, collecting wild fruits, and hunting.

As noted in chapter 1, Porritt (2007) argued that the two primordial capitals were natural and human, but I suggested that the primordial capitals would be natural, human, and *social*, specifying as well that legally recognized and defensible access to natural capital is a key requirement in a modern state. Legal rights by a population to a forested territory, the human capacity to learn and adapt, and the ability for collective action based on human norms of cooperation (Bowles and Gintis 2011) are the primordial elements that allowed for the creation of CFEs, and these processes came together in periods ranging from centuries to less than a decade.

The crucial role of "territoriality" in providing the foundation for subsistence corn farmers to operate market-oriented enterprises is demonstrated in Prisoner's Dilemma or choice games (Axelrod 1984; Ostrom 2000). In experimental choice laboratory studies, researchers have found that the study populations (mostly U.S. college students) fall into clear behavioral categories: "conditional cooperators" and "willing punishers," that is, "norm-using players" who under the "sparse institutional environment" of the laboratory can assert cooperation norms that can pressure "rational egoists," the third behavioral category, toward greater cooperation (Ostrom 2000). The role of "some social structure" and of an option to communicate has proved crucial in the laboratory evolution of cooperation. An important basis of this social structure is "territoriality," where players interact with neighbors and not just random strangers and where "success depends on how well they do in their interactions with their neighbors," which is to say, trust (Axelrod 1984, 158; Fukuyama 1996). When two individuals have a high or certain probability of future encounters, the future is said to have a "large shadow" (Axelrod 1984, 174). In Mexico, territoriality had a physical manifestation, where human intelligence and social networks were legally attached to forests, supplying a strong incentive for ongoing cooperation, becoming stronger when the incentive became robustly economic. Some Mexican forest communities had preexisting "large shadows" of territorial interaction with extremely well-defined and time-tested rules (Aguirre Beltran 1979; Bray 2008). However, other cases show that even in communities with baselines of disorganization, the combination of institutions from above that greatly reduce transaction costs and strong price incentives can create unexpected feats of cooperation around enterprise development.

To show how state policies, market incentives, territoriality, and community response were intertwined in historical processes, I will present two extensive and contrasting case studies of the historical development of two of the larger CFEs in Mexico. The two cases represent the two poles of origins for CFEs, from deep pre-Hispanic roots to "communities" formed only in the 1970s. They are San Pedro el Alto in the Southern Sierra of Oaxaca, a Zapotec indigenous community with roots in the region going back to the pre-Hispanic period, and El Balcón, in the Sierra Madre del Sur of Guerrero, formed by mestizo colonists in Guerrero in the 1960s, which had some three decades of success before recently falling victim to a climate of rural violence associated with organized crime. In the remainder of the chapter, I will expand the lens on the origins of community in Mexico to further emphasize the multiple pathways that have led to the formation of CFEs and their success or failure.

SAN PEDRO EL ALTO, ZIMATLÁN MUNICIPALITY, SOUTHERN SIERRA, OAXACA

Today, the Zapotec indigenous community of San Pedro el Alto (San Pedro) is one of the superpowers of Mexican community forestry.[4] With some 1,800 inhabitants and 350 comuneros, the community has 30,048 hectares of territory, of which around 26,000 hectares of pine forests are of high commercial quality and of which an estimated 7.5 million cubic meters are standing stock. As of 2018, it had an annual authorized volume of 100,000 cubic meters and annual timber sales of some US$4 million. It has two sawmills; wood-drying facilities; a water-bottling operation (run by women from the community); a nursery; a bus line; a gas station; a savings cooperative; and grocery, pharmacy, and stationery stores. It is also launching an ecotourism operation. All operations are administered from a three-story office building in Oaxaca City. The CFE employs some 200 workers, around 60 percent from San Pedro and 40 percent from surrounding communities; additional employment comes from the other community enterprises.

San Pedro has deep pre-Hispanic roots. Zapotec *señoríos* (minor lords paying tribute to a regional Mixtec king) have been documented in the preconquest Sierra Sur. San Pedro and other Sierra Sur communities "should be understood as historically continuous from the ancient prehispanic *señoríos* and the deeply rooted and defensive communal territoriality is an expression of an identity going back to the deep past" (Carmagnani 1993, cited in Garibay Orozco 2008, 152). Communities and territories were disrupted by the disease-fueled demographic collapse of the sixteenth century that decimated indigenous populations, leaving villages with small numbers of survivors. The survivors were

---

4. Information in this section is principally from C. Garibay Orozco (2008), supplemented by M. Rosas-Baños and R. Lara-Rodríguez (2013), by interviews with community authorities and employees in Oaxaca City, and by a two-day visit to San Pedro in December 2018.

concentrated into population centers known as *congregaciones* or Indian Republics by Spanish colonial authorities. Oral histories in San Pedro suggest that what is today a small population center in the territory, known appropriately as Pueblo Viejo (Old Town), was the precolonial settlement of San Pedro. In a December 2018 visit to San Pedro, one of several over the years, I saw proudly displayed in the municipal building a reproduction of a seventeenth-century map that shows San Pedro with boundaries similar to those of today.

Indian Republics such as San Pedro were "stable, strongly autonomous and very well-structured" (Garibay Orozco 2008, 156), and the church would become the symbol of the town. Colonial bureaucratic control took the form of the *cabildo*, a municipal government with positions such as the *gobernador, regidor, alcalde, mayor, escribano*, and *fiscal* established by the colonial government but with substantial local autonomy in setting rules over electoral processes and other aspects of daily life (Yannakakis 2008). Thus, the Indian Republics had long experience in maintaining strong degrees of autonomy in bureaucratized hierarchies, learning to negotiate between the demands of clergy and colonial authorities. By the eighteenth century, a formalized hierarchy of service positions (*cargos*) with ascent by merit and age emerged. In addition to its presence on the map in the municipal office, the existence of San Pedro el Alto is documented in 1707, early in the historical record, as a *congregación*. In that year, the community leaders requested of the colonial bureaucracy a demarcation of the "mountains and canyons that we have enjoyed quiet and pacific possession of and that our fathers and grandfathers enjoyed" (Garibay Orozco 2008, 163), indicating similar boundaries to the community territory today. A burst of prosperity in the late eighteenth century due to the production of cochineal allowed for the building of the current church. By the early nineteenth-century independence period, San Pedro had become a solid corporatist organization that defended its territory, administered the governance institutions of the cargos, and had substantial autonomy in managing its affairs. The indigenous communities of independent Mexico were the owners of their territories, but the Ley Lerdo of 1856 declared the end of these civil corporations and called for the distribution of lands as private property. These rules launched a hundred years of ambiguity in land titling, but the remoteness and poverty of the Sierra Sur meant that a private property regime never took hold there (Garibay Orozco 2008).

After the Mexican Revolution, the agrarian reform once again permitted legalized communal landholdings. However, existing ambiguities in community boundaries that had not been clarified previously now became life-and-death struggles with the possibility of legally sanctioned boundaries. There emerged in the Sierra Sur "a kind of low-intensity intercommunity warfare" (Garibay Orozco 2008, 170) that has included confrontations and killings up to recent times. The titling of agrarian properties dragged on for decades without resolution, and the Sierra Sur became a "balkanized region composed of communities that are similar to a kind of 'microstate' in a condition of recurrent belligerence and territorial tension" (142).

Despite its ancient territorial occupation and dense social capital, San Pedro, at the end of the 1940s, was a very poor, remote, Zapotec-speaking community based on subsistence agriculture and small-scale livestock raising. Its inhabitants lived in small adobe houses with dirt floors lit only by pine torches, two long days on foot from Oaxaca City. Food production was frequently uncertain, and high rates of disease and infant mortality were a part of life. Forests were used mostly for subsistence (firewood, medicinal plants, building materials for houses, wild fruits, and hunting); the sale of fatwood (resin-impregnated heartwood) was one of the few commercial uses. The community continued to be governed by the cargo system. However, after hundreds of years of little basic change, things started to transform in 1947, when the Forest Company of Oaxaca (CFO) won a concession to log the forests of the Sierra Sur. This concession marked the conversion of the forested territory of San Pedro into natural capital, as an essential natural resource in a modern industrial process. San Pedro still did not have formal legal title to its territory, but community rights to the forest and trees were nonetheless respected: the company had to convince San Pedro to sign annual contracts, as required by the concession. The CFO promised jobs, roads, and services but, most crucially for the community, support in finally getting a legal title to its land and resolution of boundary disputes with no fewer than five of its neighbors, continuing the territorial claim from 1707. San Pedro signed a contract with the CFO in 1948, and by 1954, lightning fast for processes that had lasted decades, it had a presidential resolution as a comunidad for 30,048 hectares, its endowment of natural capital. In the meantime, the CFO used its financial capital to invest in the physical capital of a sawmill, logging roads, and extensive industrial installations including dormitories for lumberjacks brought in from Michoacán. Eventually a company town in the forest grew to have a population of some two thousand. Cultural influences from the presence of so many outsiders meant a progressive loss of the Zapotec language and the disappearance or hiding of some traditional cultural practices. Roads meant that the community was now a few hours and not days from

Oaxaca City, and new merchandise and government services, including teachers and clinics, began to arrive.

The annual logging contracts included derechos de monte and set salaries for the community members who worked in the enterprise. In the paternalistic practices of the time, the derecho de monte was deposited into a fund controlled by Agrarian Reform (a predecessor of FONAFE) that the community had difficulty accessing. In the first few years, contracts were signed by the community without being questioned. But as the years went by, the community learned how to negotiate and began to exercise its "hold-up rights," a sign of the presence of new human and social capital stimulated by interaction with the private enterprise, and it demanded more benefits from CFE. It bargained for roads to the population centers within the territory and construction of schools, clinics, and housing, in a first sign of political agency and collective action related to forest issues. The community governance institutions from the colonial and independence periods endowed San Pedro with considerable social capital to augment indigenous solidarity, allowing the community to present an organized negotiating front. As workers, they were incorporated into industrial forestry in the most menial unskilled positions. However, in the annual negotiations the community members began to learn the rudiments of calculating timber volumes, converting cubic meters of logs to board feet, and creating balance statements, information that told them how much the CFO owed them, further building social and human capital. Substantial sums of money began to be deposited in the Agrarian Reform fund, but the government agency made independent decisions on how the funds should be used, occasioning frequent disputes over their use. The community learned to become aggressive in its complaints and by 1958 forced Agrarian Reform to grant it direct decision-making over the fund, in the same period when communities in Durango were demanding territorial rights and their own CFEs.

There is no evidence of divisions within the community during this period or later, and the community began to take on many of the characteristics of a labor union:

> The communal assembly maintained itself always in front, and with an astonishing continuity as the center of communal power. In the historical documents, we don't find any sign of political ruptures . . . In fact, there is a strong bloc posture in the annual renewals of the contract, in complaints to government officials, or demanding benefits for community members. They began to act like a highly solidary labor union, and in fact formed relationships with national labor

syndicates and named the new school after the Martyrs of Chicago. (Garibay Orozco 2008, 184)

With access to the Agrarian Reform fund, the Assembly bought a passenger bus to Oaxaca City, their first directly entrepreneurial activity. The community then bought logging trucks and negotiated with the CFO to be the exclusive supplier of trucking services. Eventually a few members received training as chainsaw operators. "Beginning with those first experiences in negotiation and organization, San Pedro gained confidence in themselves, which later served them in starting their own industry" (Garibay Orozco 2008, 185). Clearly, important economic incentives began to make community collective action stronger and more capable, though the community's members did not yet glimpse the option of operating the industry themselves.

The relationship between the CFO and San Pedro remained basically stable over two decades, with annual negotiations over the contract, until the end of the 1970s. By then, communities in both the Sierra Sur and the Sierra Norte of Oaxaca saw that the twenty-five-year concessions were ending and began to make more demands on the concessionaires, as detailed in chapter 2. In San Pedro, by 1979 the annual negotiations became more difficult, and by 1980 there was a sharp confrontation between San Pedro and the CFO, which included the community seizing equipment, blocking access to the forest, and causing the suspension of logging activities while a political fight continued, similar to the resistance of communities in Sierra Norte during the same period. By 1982 the twenty-six communities of the Sierra Sur that had contracts with the CFO banded together, demanding an end to the concessions. Nonetheless, in one of his last acts in office, President López Portillo renewed the concessions, as also noted in chapter 2. However, as in Sierra Norte, the communities filed a legal challenge against the renewal, which was supported in the courts, and eventually the concession renewal was overturned by the incoming president de la Madrid, the presidential intervention previously noted. The formal legal institutional and political framework of Mexico sided decisively with community-directed forestry at this historical juncture. The communities in both Sierra Sur and Sierra Norte could now begin to organize their own CFEs. However, this was not an easy task, despite the ownership of natural capital and the accumulated but inadequate other types of capital.

During the concession period, the CFO had run its enterprise using the natural capital of San Pedro's forest but providing all of the other capitals itself.

From 1980 to 1984, with the end of the relationship with the CFO, logging in San Pedro stopped entirely. Despite inheriting the physical capital of the logging roads and gaining control over its natural capital, San Pedro still lacked essential human, social, physical, and financial capital to actually operate its own CFE. In that period, many comuneros had to migrate for work, having become accustomed to regular cash income. Also in that period the DGDF began to operate in both the Sierras of Oaxaca. As discussed in chapter 2, the DGDF, under the leadership of León Jorge Castaños, was a remarkable entity operating as a boundary or linking pin organizations (Carley and Christie 2000). As noted by team member Martha Calleros, "To a large degree we functioned as an institution independent of the official structures, although obviously with some support, some representation" (ASETECO 2002, 55). To San Pedro and other communities in the Sierra Norte and the Sierra Sur, DGDF workers offered studies, training in sawmill administration, and other human capital training necessary to operate a modern forest enterprise, as well as political advising. It was not until the end of 1984, seventeen years after the first CFEs in Durango, and with the support of the DGDF team, that the community first founded its independent CFE, the San Pedro Specialized Economic Unit for Communal Logging (UEAFC). Specialized Economic Units for Communal Logging were a new legal figure that had been created by Agrarian Reform to encourage CFEs, and it entailed additional support from the formal legal institutions of Mexico. The DGDF, as a government agency, made the investment in the stock of the crucial new human capital by instructing on how to organize, administer, and finance a logging enterprise.

In 1984, the first year of independent CFE operation, the DGDF arranged for a loan of a winch and a logging truck to begin operations, with a scarce 1,000 cubic meters of volume the first year. The CFE grew slowly as its members gained experience, and within four years the CFE had achieved 5,000 cubic meters. By 1990 there was a great leap forward; San Pedro gained authorized permits for up to 40,000 cubic meters annually, bringing with it a renewed burst of relative prosperity to the community, and was able to reinvest some of its financial capital in its own sawmill at the edge of town. Most of this industrial expansion was self-financed from the logging sales. To control the new flow of financial resources, the DGDF introduced the organizational and institutional innovation of the Comisión Revisora (Review Commission), which served as a financial auditing arm of the Assembly. Parallel to initiating this service, the DGDF in 1985 promoted the UCEFO, a further investment in linking social capital. This intercommunity organization sought to jointly negotiate timber prices, provide forest technical services, and offer accounting and other business administration services to six communities in both Sierra Sur and Sierra Norte (though UCEFO stopped functioning later in the 1990s). Growing steadily and achieving a high level of success, in 2001 San Pedro was certified for good management by the Forest Stewardship Council.

Today, and after years of accompaniment by the DGDF and the NGO ASE-TECO, formed by former staff of the DGDF, San Pedro is a fully autonomous CFE. It calls on technical assistance as it feels it needs it, and the permanent staff includes at least one member of the DGDF team from the 1980s. It is a "village-level collectivist economy centered around an industrial forest enterprise" (Garibay Orozco 2008, 195). It employs more than 350 workers year round in all its industries, from both San Pedro and neighboring communities. Logging an annual volume of 100,000 cubic meters and bringing in an annual gross income of up to US$4 million a year, community members have mastered all the multiple skills necessary to operate a modern industrial forest enterprise, from chainsaws through heavy machinery and sawmill operations and all office, marketing, and representational roles; they have also rehabilitated the abandoned operations center left by the CFO in the forest. "Family incomes, relatively high for a rural area in Mexico, depend almost totally upon the community enterprise, with only a small number of families still doing subsistence agriculture" (Garibay Orozco 2008, 195). CFE members are extremely focused on economic equality within the community, reflected in a compressed salary range between the top enterprise directors and unskilled labor in the forest. Until recently, there was no direct profit-sharing, but due to higher authorized volumes the CFE has started annual profit-sharing of US$1,600 to each community member household in a recent year, equivalent to several months' salary at rural labor rates. All other profits are reinvested in the enterprise, in public works, in community governance, and in community religious festivals.

Public good benefits include potable water systems, road and street maintenance, clinics, school construction, irrigation systems for agriculture, satellite cable television for the entire community, and a magnificent municipal building. There is also a social welfare safety net, including a free clinic staffed by a full-time doctor, discounts for medicines, modest retirement pensions for community members over sixty, support for widows, a community store, and school transportation for children from outlying settlements, while the regional bus service operates essentially at cost.

An extensive calendar of religious and civil ceremonies has been paid for almost entirely from enterprise profits. The forty-six such events exhibit a tight coupling of both religious and civil ceremonies and an "enthusiasm for community exaltation," since many of the fiestas strengthen identification with the community (Garibay Orozco 2008, 199). This celebration of community is combined with staggering demands for community service. The traditional civil-religious cargo system includes thirteen positions, each of which last a year. Although periods of rest occur between each year of service, it is in the interest of the commoner to start discharging functions in their younger years. Cargos can be enormously demanding, such as the *tequitlato*, a sort of errand runner for the civil authorities, requiring one full week a month for thirteen-hour days. In addition to the cargos, multiple committees—for schools, the clinic, and special events—add to the need for participation. Serving for a year on the Review Commission, established decades before by the DGDF, gives comuneros basic training and experience in accounting and the ability to present auditing statements to the community assembly. There are also the demands of governance by Mexican agrarian law, the Comisariado and the Oversight Council, with the agrarian positions serving for periods of three years.

With all these service obligations, seldom a year passes in the life of comuneros in which they do not have varying degrees of obligation to community service. In addition to these obligations, community assemblies occur once a month and are never postponed. They can last all day Saturday until 9 or 10 at night and will continue on Sundays if there are still items to consider, and attendance is obligatory. The Assemblies are unforgiving in assigning unpopular tasks and "are so exhaustive in their discussions that they end up making public practically all of the details of community life" and constitute "a center of intense training in the social" beginning at the age of eighteen (Garibay Orozco 2008, 223, 231). Finally, all community members are required to offer fifteen days of *tequio*, community labor repairing roads and work in the forest on Saturdays. Failure to comply can include up to being dismissed from their job in the communal enterprise. In sum, C. Garibay Orozco terms San Pedro an "industrial forest community," a social justice version of the company town, and a fusion of a community and an enterprise that also implies a complete suppression of any rational egotists in the community. This degree of community collective action makes the New England township democracy so admired by de Tocqueville and Ostrom appear to be very weak tea. In addition, the deep ancestral connection with the territory is expressed in annual ceremonies organized by the enterprise to honor forest spirits (*chaneques*) that can be benevolent or malevolent (Garibay Orozco 2008). It is one the few documented cases in which pre-Columbian religious beliefs have been integrated into modern industrial forest management practices.

EL BALCÓN, MUNICIPALITY OF TÉCPAN DE GALEANA GUERRERO

El Balcón had dramatically different origins from San Pedro el Alto, at the other end of a continuum.[5] Nonetheless, until recently it had fixed capital assets of some US$4.2 million and profit margins of 20–35 percent, though, as we shall see, after decades of success it has encountered severe difficulties in recent years due to rural violence in Guerrero. However, its history until most recently is remarkable because both the territory and the "community" were of very recent origin. The community is located in the state of Guerrero on the coast north of Acapulco, in the Cordillera Costera del Sur, also known as the Costa Grande. The population of El Balcón was formed by nonindigenous mestizo refugees fleeing violence in other parts of Guerrero who had first settled in the area in 1960. They comprised a few families of subsistence corn farmers, goat herders, and seasonal workers for the private forest company Silvicultura Industrial, on whose concessioned state lands they settled. In the early 1960s, this region of Guerrero was enmeshed in terror and killing marked by interfamilial and inter-community clashes. In 1961 the landless residents of the forest concession filed an application for an ejido to be established on the company lands, trying to take advantage of the opportunity offered by the Agrarian Reform agency. When the neighboring community of Cuatro Cruces found out about the application, they feared that some of the lands they regarded as theirs were claimed and mounted an ambush, massacring twelve men from the community and leaving a clan of some thirty people, with only two adult men and women and children, all of whom fled to the lowlands of Guerrero. These settlers were not acclimated to a lowland environment and two years later resettled in the Sierra forest community of Bajitos de La Laguna. At the time, Bajitos de la Laguna's forests were under concession to a logging company, and there again the new

---

5. This section is based on Bray and L. Merino (2003); Garibay Orozco (n.d.); Torres Rojo, A. Guevara-Sanginés, and Bray (2005); and interviews with one current and one former government official and with one forest technician close to El Balcón, all in November 2018.

arrivals had the opportunity to work in industrial forestry. They participated in a strike and road blockade in 1966, precipitated by Bajitos de la Laguna seeking to manage its own CFE, in the same period as the very first CFEs emerged in Durango. Also in 1966, the settlers' ejido application from 1961 was finally granted for the area that they had fled, a sign that forest distribution was brisk in Guerrero in the same period in which large blocks of forest were being handed out in Durango and Chihuahua. However, due to the fear and uncertainty of returning, the nomadic families did not reoccupy the land until 1970, having recruited some additional families to claim the modest 2,400 hectares they had been granted. Thus, they did not gain a territory they could call their own until 1970, after a nearly ten-year diaspora from lands they had only briefly occupied as forest workers in the first place. The period of the diaspora apparently created a sense of solidarity and social capital among the families, and they lobbied the government for more land. In the early 1970s President Echeverría moved against the local timber interests, as he did in Chihuahua and Durango. He blamed the local timber industrialists for provoking a guerrilla uprising in the region led by Lucio Cabañas, cancelled logging concessions, and began handing out the forests to local communities, another example of executive power being wielded for reforms in the forest sector. In 1972 the government established the parastatal FOVIGRO, as noted in chapter 2, with an exclusive logging concession, a socioeconomic development orientation, and technical support from a Finnish international cooperation (Hinojosa Flores, Skutsch, and Mustalahti 2016), in an attempt to correct abuses of the private logging companies. While the government handed out valuable forests to local communities, the military moved to suppress the guerilla movement. The Cabañas insurgency was finally eliminated by the military in 1974, after the death of the guerrilla leader in a community in the municipality of Técpan de Galeana. Also in 1974, as a part of Echeverría's rural pacification campaign and to remove future temptations toward armed uprisings, El Balcón's petition for additional lands was awarded, comprising 19,150 hectares of rich natural capital, a more modest endowment than the forest grant in El Largo in Chihuahua in the same period but nonetheless substantial. Current territorial boundaries were formalized in 1986 in an unusual peaceful settlement of a boundary dispute by a land exchange with Cuatro Cruces. The settlement was negotiated by the peasant union the Central Independiente de Obreros Agrícolas y Campesinos (Independent Central of Agricultural Workers and Peasants; CIOAC), and El Balcón ended up with 25,565 hectares for 112 ejidatarios (Bustamante Álvarez 1996).

In 1975, only a year after gaining rights to the land, El Balcón signed its first contract with FOVIGRO for logging on its lands, an almost immediate conversion of the forested territory into natural capital. This relation would last eleven years, with yearly contracts signed until 1986, a much more compressed period of learning about industrial forestry than took place in San Pedro. FOVIGRO began putting in logging roads and implementing a management plan with the Finnish technical assistance, both further capital endowments. El Balcón ejidatarios were paid stumpage rights and hired as workers to build the roads and to do the logging. Although the CFE members had some background as forest workers, FOVIGRO expanded training in industrial forestry, a significant investment in human capital. The stumpage rights were paid into FONAFE and, due to an annual harvest of 40,000 cubic meters of pine, capital accumulated quickly. In 1978, only four years after receiving the territory and three years after beginning to learn about industrial logging, El Balcón had the confidence and entrepreneurial vision to use its accumulated financial capital in FONAFE to invest in its own logging and transportation company, delivering logwood to the FOVIGRO sawmill in Técpan, to become a Type III community, ahead of San Pedro el Alto for that period. El Balcón purchased seven trucks, two winches, and chainsaws and began logging its own forest. In a first test of community social capital, the drivers of the logging trucks began appropriating them as their own, but after a period of tension and negotiations, the community was able to assert its rights over the logging trucks and employ the drivers directly. This situation lasted from 1978 to 1986, during which El Balcón successfully managed its business of selling and transporting logs to FOVIGRO and continued to gain valuable experience in enterprise administration. For CFE governance, El Balcón did not have the tradition of the cargo system or the demands of being a municipal government, so its organizational platform and learning came entirely from Agrarian Reform, the Comisariado and Oversight Councils elected by the Assembly acting as the enterprise manager.

In the earlier years, FOVIGRO functioned substantially as a social development enterprise, there is no record of significant conflicts between it and El Balcón, and it was crucial in the launching of the CFE. However, by the 1980s, it became more abusive and corrupt, and El Balcón and other forest communities in the region began to struggle to free themselves from the concession. In 1980 El Balcón and three other ejidos organized themselves and began to demand higher prices for their timber and the fulfillment by FOVIGRO of the promises it had made to the communities. In 1985, the four ejidos formed the

Coordination of Forest Ejidos of the Costa Grande (Coordinadora de Ejidos Forestales de la Costa Grande) with fourteen communities demanding easier access to communal funds and more timely issuing of logging permits and protesting the high cost of carrying out the FMPs. Together they were investing in collective action and social capital at the regional level, evolving in 1988 into the eleven-member intercommunity organization the Hermenegildo Ejido Union (Unión de Ejidos Hermenegildo Galena; HGEU) (Bartra 2000). This collaboration gave El Balcón bridging social capital and new capacity to negotiate with outside actors. Although other powerful social actors were also emerging in the 1980s—criminal entrepreneurs in the drug trade who took over many communities and who were said to have links with various levels of government and the police—in this period, the communities in the HGEU resisted dominance by the *narcos* (Durán et al. 2011).

Thanks to the 1986 Forest Law (see chapter 2), El Balcón was able to contract directly with its own forest engineer that year, giving the CFE greater control and a higher level of services in forest management planning. Also in 1986, members of El Balcón visited the CFE of Nuevo San Juan Parangaricutiro in Michoacán; the example inspired El Balcón to recondition a used sawmill to launch its own milling operation, becoming a Type IV community. However, although they had mastered the basic skills of a logging operation, once they began to operate the entire chain of vertical integration, from logging to milling and selling sawnwood, community members quickly confronted serious deficiencies in training and experience. They ran into the kind of operational problems that San Pedro had encountered early in the 1980s. They also had problems in combining the hierarchical discipline required in an industrial enterprise with ejido ideas of equality. Accounting was faulty, managerial skills were limited, and the sawmill operation rapidly fell into debt and disorganization, with the Comisariado and Assembly being overwhelmed and unable to cope with the new level of business administration demands.

To deal with these setbacks, in 1988, the ejido made a crucial decision to delegate the direction of the enterprise to a professional manager, a former employee of FOVIGRO who had earlier worked for the DGDF. This decision resulted in what could be called entrepreneurial shock treatment. The manager renegotiated debts, hired a professional managerial and technical team, and made new investments in infrastructure. I first visited El Balcón in 1992 and recall the manager, João Bala, a Canadian of Portuguese descent, showing off the dryers and the double-trailered logging trucks he had implemented to add value, reduce costs, and increase productivity in the CFE. In both the sawmill and the forest logging operations, community members were at the beginning almost entirely excluded, since the manager felt they were not disciplined workers. This change created tensions between the manager and the community, but Bala was able to direct substantial new profits toward the community, which it used to build new houses, invest in agriculture, and engage in profit-sharing, increasing acceptance of the arrangement. As workers from the community were incorporated in subsequent years, they received higher salaries and learned to accept a more disciplined work experience. From their 1986 visit to San Juan Nuevo, the community leaders also brought the organizational innovation of the Council of Principals, established in 1989 and composed of around thirty community members, which functions as a sort of board of directors for the enterprise and for the community. The Council of Principals was elected by the community and composed of former community leaders and by younger people who had technical knowledge and displayed leadership skills. The Council of Principals represented a further accumulation of social capital; it acted as both as an organizational innovation and as an arena for building accountability and provided mechanisms for social monitoring and experience in conflict resolution, transferring the burden of enterprise leadership from the Comisariado to a new community-created institution.

In the 1990s community members once again occupied most of the positions in the enterprise and began to learn some managerial roles, though labor in the sawmill remained mostly contracted in the coastal town of Técpan, given its location 60 miles from the community. In the 1990s substantial new investments were made in the CFE, both from profits and bank loans. From 1992 to 1995, for example, the ejido invested some US$1.6 million in sawmill improvements and dryers. In 1997, El Balcón made an important technological leap forward, with a new highly automated sawmill, operated by a few highly skilled workers from the community. In this period, El Balcón also won several national prizes for various aspects of its logging enterprise. In 1998, the professional manager left, to be replaced by a new one, who had been the forester. Finally, in 2002, for the first time a member of the community, one who had received professional training supported by the community, became the general manager. El Balcón also reached export markets with its high-quality timber from its forest, and in the 1996–2001 period, it exported 40 percent to 45 percent of its production, constituting 65 percent of total sales. I had the opportunity to talk with the U.S. timber importer several times in the late 1990s, and he was enthusiastic about the quality of the timber coming out of El Balcón's natural forests and the community's ability to learn new entrepreneurial skills.

By the mid-2000s El Balcón had fixed capital assets of US$4.2 million and total sales of 3.6 million dollars, its average annual profits were around US$1 million, and its profit margins ranged from 20 percent to 35 percent. In 2005 as one example, US$612,000 was reinvested in the enterprise, and US$650,000 was transferred to the ejido for salaries for ejido authorities; pensions for senior citizens (over sixty-five), the elderly, widows, and people with disabilities; and fellowships for students for both high school and college. All workers were enrolled in Mexico's social security system and had health and accident insurance. Profits from the CFE also supported medicines for the clinic; construction of a public plaza; financing of a mezcal-producing business; and new houses for all community members with running water, toilets, and solar energy for electricity. Free transportation to Técpan and to their fields was provided, and free firewood was delivered to the homes. Great value was accorded to education, including comfortable houses built for the teachers to help retain them in the isolated area. In the 2000s during the logging season, some 180 community members worked in the CFE and another 120 employees in the sawmill as one of the largest employers in Técpan. Ten administrative positions in the sawmill are held by community members, but the rest of the laborers are from Técpan, generating local employment beyond the community. The lowest income paid to workers in 2005 was double the national average for manufacturing employment, though workers not from the community were paid substantially less, and as in San Pedro a narrow range exists among community members between the top salary and the lowest. El Balcón received FSC certification for sustainable management in 2002, but it was subsequently lost.

Garibay Orozco (n.d.) suggests that the Council of Principals displaced the Assembly as the maximum authority of the ejido and describes it as less a direct democracy than a system of notables or a Senate. He also notes, as with the Zapotec indigenous peoples of San Pedro, a strong egalitarian ideology despite members being mestizos of mixed ancestry. He attributes the egalitarianism to the members' origins as several extended families that developed

> a discourse of identity that considers the territory, the enterprise, and the community as a common patrimony, indivisible and not for sale, justified by the legend-history of the family clan that stayed together in their struggle to survive the violence, for the achievement of having achieved joint ownership of the

ejido territory. . . . [T]he story of the clan of poor peasant families that arrived in the area looking for land, that suffered the massacre of their men, that lived united during the period of diaspora, that reconquested/retook the lost land and founded the ejido, that grew in population, that built the community forest enterprise and that become comparatively rich. (16)

Despite the community's nonindigenous origins, a strong sense of communitarianism prevails, and concepts of private property are very weak. Even the houses that the community members live in belong to the community, and the only private property are the things inside the house, a deeper penetration of communalism even than in San Pedro. As Garibay Orozco notes, "the experience of the El Balcón ejido shows that peasant communities are capable of constructing astonishing development initiatives that conserve the environment and promote the social development of their populations" (19).

The case of El Balcón and UEHG in the 2000s seemed to offer compelling evidence that collective action around the forests at the community and intercommunity levels may serve not only to preserve forests and generate income but also to build civil society and ameliorate serious social conflicts and violence, at least for a period. Immediately to the south of El Balcón is the region of Aguas Blancas, where seventeen campesinos were massacred by police in June of 1995, an incident linked to forest disputes. To the north is the region of the so-called *campesinos ecológicos* and Rodolfo Montiel, a murky case that drew international attention over campesinos who were detained while protesting illegal logging under accusations of drug and arms possession but who came to be hailed as environmental heroes (Durán et al. 2011). But yet in the region of the sierra, where El Balcón and the UEHG have operated since the mid-1980s, there was relative social peace for many years. However, more recent events shattered that image.

As recently as the 2005–6 period, it appeared that the culture of cultivation of marijuana (on the lower slopes) and opium poppies (on the upper slopes) had not seriously impacted the CFE communities in the Sierra of Técpan. However, despite impressive accomplishments in collective action, community conflict has frequently been evident, unlike in San Pedro. El Balcón was not exempt from rural Guerrero cultural traditions of heavy drinking and being heavily armed and violent, with drug cultivation increasingly becoming part of the mix. In 2007 an El Balcón Comisariado was murdered, without any guilty parties or motives ever established. The murder of one community leader by another in a drunken celebration in the community in 2011 contributed to increased internal tensions. Homicidal violence continued in 2017, when an ex-Comisariado, the current Comisariado, and two ejidatarios were all murdered only a few months apart. The motives were unclear, but speculation ranged from internal conflicts to resistance to or involvement with *narcotraficantes* in the region. The community is surrounded by other communities heavily involved in the narcotics trade and where inhabitants have been displaced due to it (Ocampo Arista 2017). The background to these events suggest a period of increasing internal tensions within the ejido linked to a generational transfer of power to a younger and more educated leadership. However, the younger leadership, despite higher levels of education, did not appear to have the capacity for conciliation of conflicting interests within the community. In the mid-2000s two highly respected external leaders left the enterprise, leaving it entirely in the hands of community members.

The younger generation soon showed tensions between two different groups and was not able to maintain a steady hand in enterprise leadership. In 2009, supported by a private bank and the Guerrero State government, the leadership launched a furniture factory, but it has struggled to be profitable. Among other things, community members were not able to depend upon regular state government purchases of school furniture as in a multicommunity furniture factory in Oaxaca (see chapter 6). The policies of a compressed salary range also were reported to change: the young leaders awarded themselves higher salaries and were seen driving expensive new pickups. In the same period, and through the early 2010s, the enterprise began to have recurring financial problems, and at one point the bank credit it had gained access to was closed off. Due to the conflicts and to a resulting decline in forest management, El Balcón lost its FSC certification in 2010, though it retained certification for chain of custody. Conflicts over what was an appropriate level of timber harvesting led to a year-long suspension of logging in 2015–16. Even before the outbreaks of lethal violence in 2017, the conflicts within the ejido led to a decision to place the management of the sawmill with one group and of the furniture factory with the second group, while the ejido as a whole would continue to manage the logging enterprise, now as three interconnected enterprises. This arrangement did not survive, and due to the internal conflicts and external pressures, the CFE basically suspended operations as such in 2017. Nonetheless, El Balcón survived and prospered for nearly thirty years and merits recognition for its accomplishments. It is devoutly hoped that the community will be able to reorganize itself and reestablish its CFE in the near future.

THE ORIGINS OF "COMMUNITY" IN THE SES OF COMMUNITY FOREST ENTERPRISES

As the case studies have shown, the "community" in community forest enterprises can have very diverse origins, though in both cases state policy and market incentives interacted with "community" collective action to produce the observed outcomes. In San Pedro the community had deep pre-Hispanic roots in the same region, while in El Balcón community was formed rapidly from nomadic families of subsistence farmers and goat herders. What appears to have been a key to success is the chemistry of at least modest social capital, decided support from state policy, and the powerful economic incentive of a valuable forest resource acting on an existing group with homogenous backgrounds who have rights to a territory, whether long standing or recent. The process constitutes the "social" in the SES of Mexican community forestry. I have shown how vigorous community response has been in both cases, but the institutional framework also provided the openings for those responses to flower. In San Pedro, the 1706 community petitioners of San Pedro understood that the Spanish colonial bureaucracy could recognize Indian Republics and in the 1950s that they could get full legal title to their lands and in the 1980s that a legal and political system could give them the space to create their own CFE. In El Balcón, in the early 1960s, the nomadic farming and pastoral families knew that the Mexican government was giving land to landless farmers, and they knew enough to try and seize the opportunity, to pressure for more land in the early 1970s, and to quickly appropriate the support given to them by FOVI-GRO. Thus, the combustion for collective action of a group of people around a territory can either grow organically over centuries or can happen suddenly when an opportunity emerges. The capacity to reach higher levels of collective action around industrial logging can happen on slow or fast tracks. In the case of San Pedro, nearly three decades passed during which the commoners progressively learned the basic human capital skills of carrying out an industrial logging operation, from felling the trees, to extracting the logs and hauling them to the sawmill. During most of this period, the comuneros were apparently content to define themselves as increasingly well-organized workers, even though the forest was on their lands. It was only with the stirring against the concessions of the late 1970s and early 1980s that they began to glimpse the possibility of actually establishing their own CFE. By contrast, in a space of only some five years El Balcón had done the social learning to seize the political moment and request more land and, under the tutelage of the developmentalist parastatal FOVIGRO, quickly engaged in the collective action to mount its own logging and hauling enterprise. However, in both cases, the existing social and human capital were insufficient to successfully operate a CFE vertically integrated to a sawmill. In San Pedro, it was only with the infusion of human capital and other support from the governmental DGDF that comuneros were able to finally fully launch their CFE. In the case of El Balcón, it was the tutelage of an initially development-oriented parastatal that provided the initial enterprise incubator followed by the clearly autonomous decision to undergo the entrepreneurial shock treatment of hiring a professional manager.

Agrawal and Gibson (1999) have argued that "the community as shared norms" is a "utopian" construct that may have little relevance in real-world communities. They propose that communities are better viewed with a focus on "multiple actors within communities, the interactions or politics through which these interests emerge and different actors interact with each other, and the institutions that influence the outcomes of political processes" (640). But the perspective that communities are only collections of actors with different interests does not reflect both the preexisting social capital and the rapid accumulation of social capital when joined with a substantial market incentive that is seen in Mexican forest communities. These are indeed both traditional and recently created "communities as shared norms." Studies of the historical development of social capital leave the impression that it is something that can only be created through long social processes, as occurred in San Pedro. However, the social capital accumulated in nomadic and violence-traumatized families of El Balcón was also enough to serve as the foundation for a larger stock of social capital that was created in very short time periods, influenced by state policy and markets. As A. Krishna has noted in the case of the Rajasthan Watershed Development Program in India, "What is of importance . . . is the evidence that social capital—embodied in functioning and legitimate Users Committees—was not available ready-made to the program and that it was developed, actually quite quickly, in the course of program implementation" (Krishna 2000, 87).

These two cases only show some of the variety in the origins of community collective action around forests in Mexico. In the Rarámuri (Tarahumara) regions of Chihuahua, there are both communities concentrated by Jesuit missionaries in the eighteenth century and those that continued the highly dispersed settlement pattern traditional to the Rarámuri. For example, the predominantly Rarámuri community of Cusárare was the first forest ejido with a cooperative in the Sierra Tarahumara in the 1930s. But effective collective action around logging has always been challenging due to the dispersion of the population. In 1957 its 692 inhabitants lived in eighty-two tiny population centers (*rancherías*) scattered over more than 30,000 hectares (Lartigue 1983; Vatant 1990). As well, in Cusárare and many other Rarámuri territories, mestizos had established small settlements after the collapse of mining in the region in the nineteenth century. It was commonly the mestizos who applied for ejido status, controlled ejido governance, and eventually controlled the CFEs as they emerged over several decades in the twentieth century, including many of the ones promoted by the CCIT. In these cases, therefore, the "community" at the time of the birth of ejido was fractured along ethnic lines, and collective action around organizing the forest enterprise was undertaken mostly by mestizos, resulting in the Rarámuri being relegated to second-class citizenship. This region represents what is actually one of the few clusters of CFEs that are not "communities of shared norms," with resulting issues of social justice and equity.

In the indigenous region of Oaxaca, a very different dynamic took place. We have already seen the process of colonial reorganization in the case of San Pedro and the Indian Republics. The proliferation of Indian Republics in Oaxaca was the historical antecedent for the proliferation of municipios in nineteenth century and in contemporary Oaxaca, with people concentrated in nucleated villages. By the end of the eighteenth century, there were 1,466 Indian Republics in Oaxaca (Carmagnani 1993, 57). In the nineteenth century Bailón Corres noted the republics' "great capacity to use the legal framework and to defend themselves in disputes over the maintenance of their local autonomy and the control of their resources" (1999, 122). In the mestizo regions of Mexico, the extensive haciendas of the Porfiriato had reduced most Mexican peasants to landless status as peons on the haciendas or as seminomadic farmers and goat herders as was the case for the ancestors of El Balcón in Guerrero. For these populations, the agrarian reform provided their first experience in self-governance and with institutions and organizations imposed from above but which nonetheless allowed them to begin to engage in collective action around a territory. The magnitude of the historical transition from rulelessness to rules in these mestizo communities was captured by Simpson in the 1930s:

> The new ejidatarios are at least men with something to plan for, something to fight over, and not slaves and peons, landless, hopeless, and helpless. Out of

competition and conflict there may come integration, unity and order; out of the sodden impotence of peons tied in debt slavery to the hacienda, the only issue possible was stagnation and death. . . . The giving of land to the rural communities of Mexico has released energies long latent and aroused hopes long dormant. These energies need to be directed and these hopes enchanneled. (1937, 352)

Some of these hopes for the ejido sector may not have ever been fully realized, but in an important number of forest communities, energies and hopes around forests were eventually "enchanneled" into, at least in many cases, disciplined, democratic, and peaceful communities. Even in decidedly unpeaceful communities such as El Balcón, the agrarian reform unleashed community entrepreneurial energies that were successful for decades. The "institutional make-up of the nation" in rural areas was provided in a national template by the agrarian reform process. These institutions allowed, over time, for the collective action by communities to become more effective and develop into the innovative organizations of CFEs. Communities always showed vigor in defending their interests, even if the entrepreneurial vision was not always present. In Cusárare in the 1940s the Rarámuri protested the logging "cooperative" that operated as a private enterprise by the government and the mestizos that abused them as workers and demanded reforms, but they did not achieve control of the CFE for decades. In Durango in the 1950s, communities, assisted by advisors, rose up against the abusive contracts imposed upon them by private enterprises and actually launched their CFEs, as we saw in chapter 2. But only slowly and painfully, and not in all cases, did these protests get channeled into the construction of autonomous CFEs. As well, many traditional communities, even if they were capable of occasional collective action in defense of their interests, were also commonly riven by internal conflicts and mistrust. For example, M. Kearney (1986) has noted about Santa Catarina Ixtepeji in the Sierra Norte of Oaxaca the "generalized distrust of fellow townspeople" (44) in the community in the 1960s. It appears that the need for persistent collective action around CFEs helped ameliorate that traditional mistrust.

It was the agrarian governance institutions and organizations that provided the initial framework for the constitutional, collective, and operational rules of the historically emergent CFEs and bestowed upon them a bundle of rights. However, some of the rights, such as the rights of management and withdrawal, only emerged through interaction between top-down policies and community struggles. This process constituted a massive project in social learning for the varied "communities" in the forests of Mexico and beyond. The cabinet-level agency that directed this process, the Secretaría de Reforma Agraria (Secretary of Agrarian Reform; SRA), is usually correctly characterized as heavily paternalistic and focused on control of the rural populations in the first decades of the agrarian reform process. But it should be recognized that the agency was also an important vehicle for social learning and constituted a kind of extension service for community organization for a system of governance that has become a fully accepted part of rural culture. It is true that communities can be riven by competing interests, as A. Agrawal and C. C. Gibson (1999) suggest, but some of them can also consistently overcome these conflicts, for sustained periods of time if not permanently, to achieve ongoing consensus about major community goals.

CHAPTER 5

THE PERMANENT TENSION

Forms of CFE Self-Organization and Community and State Failure

I n chapter 4 we saw how the incentives for collective action of two particular communities emerged from ownership of a forest territory, the supply of governance institutions and organizations, and the five capitals that these communities were able to mobilize. Mexico's institutional framework provided the context for actors to engage in creating CFEs. However, beyond forest management, evidence suggests a strong baseline tendency of all territorially based communities in Mexico to engage in collective action around rule setting and norm creation in maintaining the common property. From a national survey Robles Berlanga (2012) found that the use of rules to govern access to the community common property is widespread, with 70 percent of agrarian properties having them. Seventy-eight percent of those surveyed reported adherence to those rules, and 84 percent reported no disagreements on how the common property was used. Specifically, of forest ejidos surveyed (defined only as those ejidos that have forests but are not necessarily CFEs), 56 percent confirmed enforced sanctions. The existence of the common property seems to provide an incentive for rule making that is only sharpened when a valuable common property forestry is owned. However, as I will discuss in the second part of this chapter, Mexican CFEs can also undergo affliction by elite capture, corruption, violence, and organized crime when institutions and organizations fail. Just as states and markets can fail, so can communities. However, in many cases community failure is principally due to the failure of the state to provide basic institutions and organizations of law enforcement or to provide the degree of forest extension efforts present in the 1970s and 1980s.

The strong community governance that emerged did not in many cases have to resolve numerous knotty dilemmas of constitutional, collective, and operational rules, because many of these rules had already been established from above. The emerging CFEs also did not have to spend time developing rules around the forest harvest, since the rules came from above in the form of environmental and forest laws and regulations. Responding to the rules from above, community governance, enterprise governance, and forest governance became fused together under the direction of the community Assembly and elected authorities. Complementing the rules from above, norms of cooperation in forest communities both emerged from the distant past and had much more recent origins. In both cases, the norms facilitate the "conditional cooperators" and "willing punishers" to dominate "rational egoists" and eliminate or control free-riding in collective action, since successful communities "can initially draw upon locally evolved norms of reciprocity and trustworthiness and the likely presence of local leaders in most community settings" (Ostrom 2000, 149).

In Mexican CFEs, these norms of cooperation and democratic governance can undergo strengthening by three intertwined dimensions, two of which were almost entirely imposed from above by the state as a facilitative political regime. The first two dimensions are the agrarian community governance platform, used or adapted to administer a market-oriented enterprise with various stages of industrial vertical integration, and the timber harvest rules, defined in the environmental and forest laws and regulations. The third dimension, the rules and organizational models adapted by the CFEs as well as rules governing the territory outside the commercial forest, is where community rule making and self-governance allow members of community forestry to be entrepreneurial protagonists, as shown in the conceptual framework in chapter 1. However, even in this case the state frequently provided the initial organizational models, as we saw with figure 4 in chapter 2. Aiming for a comprehensive view in this chapter, I will review the functioning of the agrarian governance platform and timber harvest rules as institutions that came from above and the evolution of different organizational models that emerged from both above and below in the process of the rapid emergence of industrial vertical integration, as communities strained to become that most unusual enterprise, a CFE (see chapter 6). Finally, I will discuss cases in which governance institutions and collective action around the CFE falter and in which the evolving SES has failed to provide effective community, enterprise, and forest governance, due to both community and state failures in providing robust institutions and organizations for the rule of law.

CFE GOVERNANCE INSTITUTIONS AND ORGANIZATIONAL MODELS

As noted earlier, the origins of common property governance in Mexico are not "lost in time" but have well-documented roots in the political turbulence of early twentieth-century Mexico. The basic set of institutions and organizational templates for community governance are found in the agrarian legislation that originally evolved in the 1920s and 1930s, with significant modifications in the constitutional reform of 1992. The agrarian legislation has remained the most basic governance framework for creating and perpetuating the CFE sector in Mexico (Bray 2013). The amended Agrarian Law of 1992 retains many of the features of agrarian governance established after the Mexican Revolution, while expanding and consolidating property rights over the territory. The basic organizational structure is provided by the Assembly of all legal community members in both ejidos and communities (see figure 8). The Assembly elects the Comisariado Ejidal (Ejido Supervisory Body) or Comisariado de Bienes Comunales (Indigenous Community Supervisory Body), and each is composed of a president, secretary, and treasurer. A second governing organ, designed to provide checks and balances, is the Oversight Council composed of a president and two assistants (Article 36, Ley Agraria).

This organizational template has extensive rules around its operation in the agrarian law. For example, legal members of the community are defined as those household heads on the list officially maintained by the Secretary of Agriculture. Each legal member of the community has a vote, and voting can be either by consensus or majority rule, at community discretion. Elections to office are required every three years, though community leaders can be replaced earlier for nonperformance. Assemblies are legally required to meet twice a year, but well-organized forest communities typically meet monthly or even more frequently depending on needs. Assemblies also typically establish multiple committees to carry out civic duties from establishing potable water systems and clinics to restoring churches, another area where they have rule-making autonomy. These positions are typically unsalaried, and the people who hold them are normally without skills in management or forestry. If it is a community with a traditional

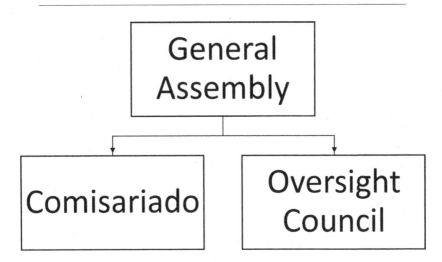

FIGURE 8. Community governance template from Mexican agrarian law.

cargo system, as in San Pedro, then the agrarian organization and rules have been incorporated into the cargo system, a form of "political syncretism" (A. Mathews, personal communication, April 12, 2015) that fuses colonial and agrarian reform governance institutions. These institutions and organizations were developed to govern a human community and a territory. They always had a clear natural-resource governance dimension, since the agrarian law calls for a zoning of land uses into agricultural, pasture, and forestry among other uses. This permitted the territorial governance platform to become the base for forest enterprise governance.

Although frequently corrupt historically and manipulated by caciques and external actors, these imposed forms of community political organization nonetheless gave peasants experience in leadership and in negotiating issues with external authorities, and some percentage of communities evolved toward a genuinely participatory village democracy (Fox 1996). Collective action around valuable forests appears to increase the odds for participatory rural democracy. However, the agrarian rules and organizations were not designed to meet the challenges of managing industrial forestry, an activity not contemplated in the agrarian legislation and with little relation to traditional indigenous or local knowledge of forests (Klooster 2002). Mexican communities, whether they traditionally had relationships with forests or were strictly agricultural communities, historically had no prior knowledge of scientific forestry or industrial forest management. Thus, the imposition from above of scientific forestry principles and silvicultural practices as embodied in the legally required FMPs greatly lowered transaction costs in the transition to organizing CFEs.

As noted previously, one central argument of common property theory is that "if a group of users is going to harvest from a resource over the long run they must devise rules related to how much, when and how different products are to be harvested" (Ostrom 2005, 262). Mexican forest communities have only limited participation in the rules of the harvest. "How much" is determined by technical specifications in Mexican forestry and environmental laws, with community rule-making autonomy lying only in the decision to possibly harvest less than the authorized volume. "When" is more relevant for plants and animals with reproductive cycles, so for forests that are growing most of the time, the "when" is determined significantly by rainfall regimes and the quality of logging roads, rather than any community rules. The "how" in a forest management sense also gets shaped heavily by environmental laws that require particular preharvest, harvest, and postharvest practices. The "how" in terms of enterprise organization models was also frequently and originally provided by government agencies, but ongoing organizational innovations provided a significant space for self-organization and community decision-making over enterprise organization. This space aside, the fact remains that much of the "how much, when, and how" of forest management is not within the realm of the constitutional, collective, and operational choice of the communities. From the time that CFEs first emerged, they have been constrained in this choice arena and have had to adapt to the absence of autonomy in most forest-harvesting decisions. Consequently, in order to understand the institutions and the organizations that govern Mexican CFEs, we must understand the constraints placed upon them by Mexican forest and environmental laws and the centrality of these rules to forest management decisions. These rules were also reviewed in chapter 3, but here I will go into more detail.

THE RULES OF FOREST HARVESTS IN MEXICO

There are three levels of rules and regulation of commercial forest extraction in Mexico, for both timber and non-timber forest products (Bray and Duran-Medina 2014):

- The LGDFS from 2003, modified in 2008 and 2013. As noted in chapter 3, a new Forest Law was promulgated in 2018, but will not be analyzed here.
- The Reglamento of the LGDFS, composed of 178 articles in forty-four pages, from 2005.
- The NOM 152-SEMARNAT-2006 (hereafter, NOM-152), which establishes the guidelines, criteria, and specifications of the contents of the FMPs approved in 2008 and is composed of twenty-four pages of detailed instructions.

The LGDFS, its regulations, and the NOM-152 together call for a systematic and detailed process for the development of the FMPs. FMPs must be carried out by university-trained foresters who are listed in official government registries. In a notable percentage of cases today, the forester may be from the forest community, increasing trust in the process. The process of planning for forest harvesting begins with a forest inventory, which requires extensive data collection and use of remote sensing and Geographic Information Systems (GIS). The inventory's creation and stand selection are done prior to seeking authorization from SEMARNAT, as part of the proposed FMP. In most cases, teams of community members carry out the data collection for the inventories under the supervision of the forester. The inventory methods required in the NOM-152 are highly detailed, using systematic or random sampling strategies with sample parcels of 500 to 1,000 square meters. After the inventory, the forest engineer's office then conducts the stand selection based on criteria such as soil, climate, slope, topography, microwatersheds, species composition, and other vegetation characteristics. Orthophotos and Google Earth images are also used for harvesting planning. Stand selection is checked with field observations, and stands and substands (*subrodales*) are classified as to quality. In most communities, the FMPs undergo discussion with the community Assembly before their presentation to SEMARNAT, so there is a degree of community participation, constrained by the technical requirements. The same general process is followed for all silvicultural regimes, though the nature of the inventory and some other steps change substantially in how they are implemented between uneven-aged and even-aged systems (see chapter 7 for more specifics on silvicultural practices and forest structure and composition).

Once the stands are determined, the proposed FMP is submitted for the review and eventual authorization by SEMARNAT. Once authorization is received, the trees are then marked for harvesting. Environmental dimensions of the regulations lay out rules for road and skid trail planning, felling of defective trees, and proper identification of commercial species before cutting. Other requirements to reduce environmental impacts include twenty-meter riparian buffer zones and no logging in forests of high conservation value, as well as measures to protect species listed in Mexico's threatened and endangered species legislation. Logging on steep slopes is also prohibited, but due to the rugged terrain in many community forest areas, logging above the maximum slope is commonly permitted, with erosion mitigation achieved through leaving chopped slash in contours. Directional felling and low-impact logging practices are also frequently used, though not explicitly called for in the regulations. In northern Mexico, low-impact and climate-friendly logging includes the use of animal traction for extraction (Bray et al. 2016). In logging of tropical forests, further documentation and mitigation of environmental impacts are required. The resulting FMPs usually run to two thick volumes with hundreds of pages of description and statistical detail. SEMARNAT and PROFEPA carry out monitoring and sanctions of the FMPs or enforcement of the regulatory framework. The extensive regulations include a significant number of elements of what is referred to elsewhere as Improved Forest Management (IFM) or Reduced Impact Logging (RIL) (Griscom and Cortez 2013; Putz et al. 2008; the environmental and ecological implications of silvicultural practices will be further examined in chapter 7). The single most important figure to emerge from this mass of data is the authorized logging volume, calculated over the ten-year period for which FMPs are normally authorized, in theory representing a sustained-yield harvest. It is common that the complete authorized volume for a logging year is not harvested for a variety of reasons, from late-arriving permits, to rains, to disorganization. However, as will be explored further in chapter 7, it is also common that communities make conservation-oriented decisions to not harvest the entire authorized volume.

Logging regulations appear to be generally followed in Type III, Type IV, and Type V communities, since they are actively involved in the extraction process. In Type II communities there is likely a wide range of compliance, from contractors who do not respect the FMP to communities who closely monitor contractor performance. Both formal written and informal community rules can and frequently do buttress the official regulations, and in well-organized communities self-monitoring and internal sanctions are more important than PROFEPA enforcement. The regulatory framework does have at least some teeth, and infractions can be documented and sanctions imposed by PROFEPA in cases of substantial noncompliance with the management plan, as in the case of El Balcón. Thus, although most rules are set externally, they have been largely internalized and depend heavily on self-policing by the communities, measures the communities commonly see in their own self-interest. Since the communities have operated within various iterations of these regulations for decades, it is likely that these measures have become incorporated into community norms, as a component of a new community culture of industrial forestry. Although externally imposed, forest regulation in Mexico needs to be seen in a context of substantial devolution or decentralization of rights over the forest resource (Hajjar, Kozak, and Innes 2012), as noted in chapter 3.

SELF-GOVERNANCE OF COMMUNITY FOREST ENTERPRISES

In addition to the harvest rules, and as seen in chapter 2, in many cases the basic organizational model for how to modify the community governance organization to adapt to enterprise administration also came from above, though not with the force of law. However, once the template was provided, the communities exercised constitutional, collective, and operational choice in how production is organized, in decisions to vertically integrate or not, in how to organize the labor force, in choosing a buyer and sale price, and in policies for wage and profit sharing, that is, substantial rule-making autonomy and self-governance of their CFEs. As Garibay Orozco notes, organizing a CFE requires "the centralized control of the territory, of labor, of monetary capital, combined with a sophisticated system of control and equilibrium," none of which are anticipated as responsibilities of Assemblies and Comisariados in agrarian law (2008, 439).

Because CFE self-governance is the major arena of community rule-making, it is important to understand its varying forms. Three different major forms of CFE governance have been proposed, "community," "work groups," and "individual" (Antinori and Rausser 2010). Most CFEs first emerged as fully "community" CFEs, with work group and individual options emerging as responses to perceived problems with the community CFEs over decades. However, there is also wide variability and organizational complexity in the community form, depending on where day-to-day managerial control is located and the degree of diversification of the forest enterprise.

VARIETIES OF THE COMMUNITY MODEL OF CFE ORGANIZATION

The figures that follow map out the organizational and institutional innovations of community governance institutions adapted to the administrative demands of a market-oriented enterprise, drawing on both models suggested

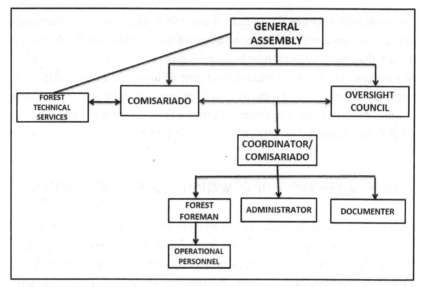

FIGURE 9. Comisariado model of CFE governance.

by government extension agents and community innovation. Figure 9 shows the Comisariado Model of CFE Management, typical of Type II (stumpage) and Type III (extraction) communities and small Type IV (sawmill) communities.

The empresas ejidales forestales promoted by FONAFE in the 1970s involved a model of the organizational enterprise, though this has received little documentation. Later in the 1970s, the DGDF developed several models for the organization of the Raw Materials Ejido Production Units that it promoted in Puebla and elsewhere (see figure 4 in chapter 2). Figure 9 shows a modification of the most basic of these models. The Assembly of the community is the maximum authority and, as noted above, the Assembly votes for the Comisariado (president, secretary, and treasurer) and Oversight Council (president and two *suplentes*, or second-in-commands), with the members of the Oversight Council frequently being assigned diverse management tasks not necessarily related to oversight. In this model, the agrarian governance form stays the same, but the Comisariado now has new responsibilities as manager of the enterprise, and a new legally required technical advisor in the form of the forest engineer, who creates the FMP; both the Comisariado and forest engineer are under the supervision of the Assembly. The Comisariado may designate someone in the community as the coordinator of the CFE, but, most commonly, the president of the Comisariado assumes this role. A strong participatory relationship between the forester and the Assembly, mediated by the Comisariado, has been found to be key for positive results in forest management practices and good forest ecological conditions (Antinori and Rausser 2007). The coordinator then supervises the operational aspects of the logging and associated processes. The principle roles here are the jefes de monte, who directly supervise logging operations in the forest; the administrator or accountant; and the documenter, who handles the voluminous paperwork involved. There are multiple variations of this framework, but this simple structure remains the case in the majority of Type II and Type III communities, and in small Type IV communities. In indigenous Oaxaca, from the beginning this structure was complemented by the Consejo de Caracterizados (Council of Elders), composed of individuals who had passed through all of the cargo system or who have otherwise won prestige within the community, an organizational and cultural legacy repurposed as a sort of "community Senate" (F. Chapela y Mendoza, personal communication, April 12, 2015). The Council of Elders historically had the charge of resolving community conflicts but had little expertise in resolving conflicts that required business skills.

In all organizational forms, the FMP developed by the professional forester and submitted to SEMARNAT is subject to approval by the Assembly. In earlier periods in Type II communities, jefes de monte and all or most forest workers were direct employees of outside contractors, and the community had little control. This period is frequently referred to in the literature as *rentismo*, whereby the communities just "rent" their forest and an outside contractor does all harvesting activities with little or no community supervision. This "classical" form of rentismo is relatively less present today, though the term is still used inappropriately for all Type II communities. For example, in a sample of thirty-one logging communities in Michoacán and Durango the jefe de monte position was paid by the community in thirty of the cases, in principal giving the community a significant degree of control over the extraction process. However, in a sample of forty-four communities in Oaxaca, in eleven of the fourteen Type II communities, the salary was paid by the contractor, suggesting a lower degree of community control (Antinori and Rausser 2010). In the few analyses of Type II and Type III communities available, a wide variety of degrees of control over the process has been documented. At one end of the continuum, in the Chihuahua municipality of Guadalupe y Calvo, a classic rentista model is still common, where outside contractors control the entire extraction process (Pérez-Cirera 2004). Moreover, Oaxaca and Michoacán have Type II communities in which "community members rarely go to the forest, have no vigilance program, and merely await payment following the timber harvest" (Barsimantov 2010, 59). On the other hand, a more recent report from Chihuahua suggests that many Type IIs in the state provide the labor for the contractor (Estrada 2018), and in Michoacán, the Type II community of Opopeo provides the timber for thriving village chair-making shops, after the timber is processed in noncommunity sawmills, and another Type II community in Michoacán is deeply involved in all aspects of forest management with the support of a local NGO (Navia-Antezana, Marín-Togo, and Cumana-Navia 2018). In the Unidad de Administración Forestal (Forest Administration Unit; UAF) Santiago Papasquiaro in Durango, a technical service provider, there are three Type II communities and two Type IIIs with forest certification, suggesting high degrees of community control (interview with F. Salazar, February 6, 2019). In another aspect of variation in management practices, it is common in Oaxaca at all levels of vertical integration that CFE management is incorporated into the traditional civil-religious cargo system, with half of the CFEs in a forty-two-community sample managing their forests using the cargo system (Antinori 2005).

Type III communities can exhibit two stages of vertical integration. In the first stage, a CFE may acquire tractors, skidders, or truck-mounted winches (*grúas*), the equipment necessary to bring the felled tree out to a road where it can be loaded onto a truck. In the second phase, the community acquires logging trucks for delivery to a sawmill. Commonly, the logging trucks are not directly owned by the CFE but are contracted with community members, passing the acquisition and maintenance costs on to them. Although there are few case studies of Type III communities, the higher level of vertical integration entailed strongly suggests more involvement in forest management, and it is likely that in most of those communities, most of the workers and the forest foreman are from the community (see Navia-Antezana et al. 2018 for an example of a well-functioning Type III community). In February 2019 I visited the Type III community of Los Altares in Durango, where I was told that it had acquired its first truck-mounted winch from the parastatal PROFORMEX in 1982, an example of the development functions of the parastatals. It has a small packing box factory that it is currently renting out, and it has recently tripled its authorized volume to 22,094 cubic meters leading to a 2018 profit sharing of nearly US$7,000 in addition to sixty full-time jobs both from the community and neighboring communities. It received FSC certification for the first time in September 2018 and is an example of a highly successful Type III CFE that has operated for decades delivering benefits to its community, despite pressure from organized crime in the region. For both Type IIs and IIIs, there appears to be some correlation between degree of vertical integration and size of commercial forest: a study of CFEs in the ten most important forest states found that Type II communities had an average of 922 hectares and Type III communities an average of 1,533 hectares of commercial forest (Bray et al. 2007)

In figure 10, a more formal enterprise governance structure emerges, an organizational framework that facilitates collective action around a relatively sophisticated industrial enterprise including a sawmill and more specialized managerial control. The operation of a more vertically integrated CFE brings many complicated administrative issues to the table that can be beyond the competency of the Assembly, and the administrative function of the Consultative Council (Consejo Consultivo), using various names, has become a common

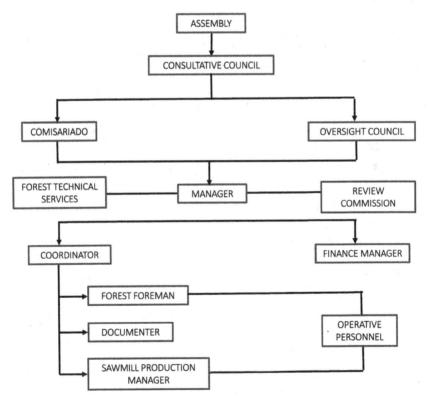

FIGURE 10. Manager model of CFE governance (Type IV sawmill communities).

practice. San Juan Nuevo Parangaricutiro in Michoacán, one of Mexico's best-known CFEs (Bray 2002), was directly inspired by the Council of Elders in Oaxaca to establish a Consultative Council and in turn inspired a similar administrative body in El Balcón (see chapter 4). However, it modernized its function by giving it a permanent managerial role and explicit responsibilities (Garibay Orozco 2008), an organizational innovation that allowed for more efficient and better-informed decision-making, and similar councils emerged in other CFEs in a process of social learning.

Below this level of shared supervisory power, there may be two variations. In Variation I, the Comisariado may continue to directly administer the CFE, even though the task may have become complex and demanding. This structural variant is frequently regarded as a bottleneck to more efficient CFEs, because the lack of professional expertise can create problems. In Variation II, shown in figure 10, a professional manager not subject to three-year elected terms is introduced. Hiring a manager was frequently a major leap of faith by communities, as we saw in the case of El Balcón, because it meant trusting an outsider. Today, it is common for the professional manager role to be filled by someone with professional degrees from the community. Thus, Variation II may be regarded as an increase in both human and social capital that can make the CFE more efficient and competitive in the marketplace. The Review Committee, as we saw in the case study of San Pedro el Alto, was an organizational innovation introduced by the DGDF in Oaxaca to provide for a greater degree of community control and as a barrier to corruption (ASETECO 2003). A variety of organizational models under the manager can exist, but it generally includes more specialized accounting functions and the new role of the sawmill manager, in addition to the forest foreman and documenter and other roles. Type IV communities generally have larger commercial forests, an average of 3,503 hectares (Bray et al. 2007), but Ejido Santa María de los Ángeles in Michoacán operates a sawmill with only 436 hectares of commercial forest (Navia Antezana, Marín-Togo, and Cumana-Navia 2018).

In Figure 11, we see the organizational framework of a diversified finished-products CFE, or Type V. Type V communities have more value-added processing that may include wood dryers and furniture, as well as factories for packing boxes, molding, and plywood. In general, more integrated communities have more permanent employment positions in administration, documentation, accounting, and technical services (Antinori 2005). In this organizational

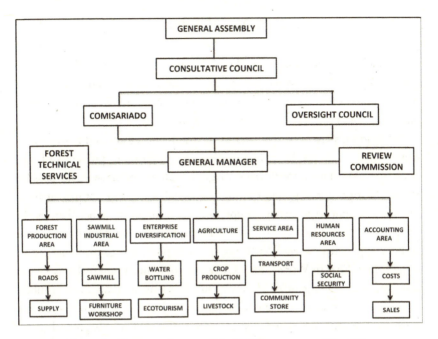

FIGURE 11. Diversified forest products model of CFE governance (Type V).

framework, the administrative hierarchy established in sawmill communities is consolidated and expanded. There remains a Consultative Council, which functions as a particularly engaged board of directors. The general manager may be responsible for the timber industry and all community enterprises, while more specialized managers administer other forest-based enterprises (water bottling, ecotourism) in addition to nonforest-based enterprises (community stores, agricultural projects, gas stations), with all of the managers reporting to both the Consultative Council and the General Assembly.

Finally, the proposed Type VI communities, as mentioned above, refers to those who have specialized in NTFPs, most commonly ecotourism, water bottling, and pine resin, as well as strict conservation under the Payment for Hydrological Services program (see chapter 7) (Hernández et al. 2015). With the exception of Type VI communities, vertical integration suggests higher transaction costs with each level, successively higher levels of social and human capital, the elaboration of more detailed rules to govern the enterprise, and greater organizational complexity.

WORK GROUPS

The models presented above are all modifications of the "one unified CFE" or community model. However, community political factions and/or corruption may undermine the community models, and in the early 1990s an alternative emerged, work groups (*grupos de trabajo*), as shown in figure 12.

Work groups are both the product of top-down reform and a grassroots-driven effort at enterprise reorganization as a response to problems (Taylor 2000, 2005; Wilshusen 2005). The work group option was made legally possible by Article 105 in the Agrarian Law of 1992, opening the door for management of common pool resources by groups within a community rather than by the entire community (Taylor 2005). It was adopted in many different parts of the country apparently independently and is an important organizational evolution of the CFEs. In Work Group CFEs the community enterprise as such is dissolved into self-organized groups or subcoalitions of community members who normally divide up the annual authorized volume on a proportional basis. Each one then operates as an independent subcommunal private group, logging

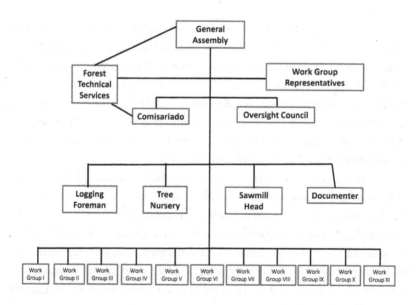

FIGURE 12. Work group model of CFE governance (chart adapted from Wilshusen 2005).

its assigned volume. There are no figures on the number of work groups in the country, but they are common in Quintana Roo, Durango, and Chihuahua. In a sample of thirty-eight communities in Chihuahua, V. Pérez-Cirera (2004) found that seven of them were organized as work groups. P. R. Wilshusen (2005) studied the emergence of the work group in the Quintana Roo communities of Caobas and Petcacab, and figure 12 is based on Petcacab. In both cases, work groups emerged to "overcome chronic infighting over administration of communal funds" (158). In Petcacab there are eleven work groups represented, further subdivided into a total of eighteen forest areas. Work groups were reported to have achieved a better distribution of income, notably higher incomes per ejidatario, and ended centralized corruption in the CFE. Work groups in Petcacab required the elaboration of new rules for the distribution of rights and responsibilities including several years of experimentation, leading to a "complex rule system that emphasized group separation rather than collaboration" (166). Timber volumes were apportioned to individual community members, and while most members pooled their shares to be managed by their group, some young community members sold their shares. These transactions created an internal market, turning the timber volume into tradable shares, which included selling futures. Creating a market for timber volume shares led to increasing inequality in the ejido, causing three members to control 29 percent of Petcacab's timber volume and reinforcing existing power inequalities. As well, the work groups reorganized community governance, becoming a "type of representative governing council that supplanted the ejido assembly" and the Comisariado (173).

In the Durango community of Canelas, eleven groups informally mapped the forest and essentially devolved rights to individual forest parcel holders, resulting in the fencing of individual plots in some cases. Nonetheless, logging still operates under a single management plan for the entire territory, making this division of rights similar to that of San Juan Nuevo Parangaricutiro (Taylor 2000, 2005). Ejido Pueblo Nuevo in Durango—comprising 243,349 total hectares, 73,921 hectares of forest, 1,688 legal community members, and 45 population centers—had six work groups as of 2019 (G. Chapela y Mendoza and Meraz n.d.; F. Salazar, personal communication). Work groups complicate representation and the administration of the FMP, since the forester and external organizations have to deal with the leaders of each group, not just one CFE manager. Work groups can also reproduce on a small scale problems of interpersonal local politics and lead to privatization of the forest, and they usually foster substantially less investment in community projects and in assets like a sawmill. However, in a case such as Pueblo Nuevo, with its large territory and dispersed population, it seems evident that the transaction costs of maintaining a unified CFE are just too high. As another indication, Pueblo Nuevo, actually a Type III community, has the only functioning sawmill operated by a private person. Although commonly interpreted as failures of community collective action, work groups in situations such as Pueblo Nuevo and other CFEs with large territories and multiple population centers are organizational innovations that may be the only practical alternative. As P. L. Taylor (2005, 145) suggests, work groups can perhaps best be seen as "one of several organizational options encouraging broad participation in forestry" and with work groups individual incomes can go up.

INDIVIDUAL ORGANIZATION

The final major organizational innovation to be discussed is the individual form of production, where forests have been illegally parceled out to individual community members. This can happen in communities without FMPs, and in these cases harvests are illegal. But in the case of CFEs, the individuals all must conform to the management plan in their treatment of their parcel, so some degree of community-level coordination remains. Most commonly, individual organization occurs with small forests but also depends on other aspects of organization. An unusual case of individualized parcels, which has nonetheless mounted one of the most successful CFEs in Mexico, is the previously mentioned San Juan Nuevo Parangaricutiro in Michoacán. Due to its peculiar history, in San Juan Nuevo communal production was erected on the basis of individually appropriated forest parcels. In the 1940s, before community logging became an option, the forests were appropriated by individual community members for pine resin extraction, effectively privatizing it. However, beginning in the early 1980s, communally oriented leaders negotiated with the private landholders to get them to agree to follow a community FMP and allow the communal enterprise to log on their lands in exchange for the payment of stumpage fees, which can be considerable amounts of money (Bray 2002; Garibay 2008). The production of pine resin remains privately appropriated, and the parcels are treated as de facto private property. A more typical example of individual appropriation is the Quintana Roo community of Cuauhtémoc. This community harvests only lesser-known tropical species under a management plan, but the entire forest has been internally and informally parceled out among the community members. Each community member individually harvests the timber on his or her land with the authorized volume proportionally divided (Bray and Merino-Pérez 2004).

Given the variety of communal, work group, and individual rights over forest resources, collective and operational decision-making around rules such as access to employment, wage levels, and profit distribution may occur at different levels. For example, in the community model in the Michoacán-Durango study, the Assembly made the decision about profit sharing in 62 percent of the cases and the Comisariado in 33 percent. In work group communities, the Assembly made the decision in 60 percent of the cases while the decision on profit rules was pushed down to the group leader or work group assembly (20 percent each) in the remainder of the cases (Antinori and Rausser 2010). There is some relationship between forms of organization and vertical integration in that the less centrally organized forms do not normally appear to have the capacity to engage in collective action for sawmill management. The Michoacán-Durango study of forty-one communities found that no individual form of organization had a sawmill and only one work group had a sawmill. The community form of organization predominated in this sample (twenty-six out of the forty-one), while the remaining communities were relatively evenly divided between the work group and individual forms and were concentrated in Type II and III communities. Thus, there is a general correspondence between the form of governance and community decision to vertically integrate into "downstream" activities. The economic dimensions of decisions for vertical integration and deintegration will be reviewed in chapter 6.

Governance institutions in Mexican CFEs are suffering strains due to emigration. Many able-bodied men leaving means fewer of them to staff the many positions in municipal government, agrarian community governance, cargo systems (where they exist), and CFE administration. In cargo systems, communities have attempted to adapt to these problems by eliminating some lower-level cargos, extending the age of retirement from service, reducing projects requiring community labor, and paying modest salaries for some positions when they can afford it. Communities have also required emigrants to come back to perform community service positions or to pay someone else to carry out the service, but considerable tensions can exist in some communities between those who have stayed and those who have left but want to retain rights in the community

(Robson 2019; Robson and Klooster 2018; Sosa Pérez and Robson 2018). This variety of governance adaptation experienced in response to migration indicates there is no one right way to manage a common property forest resource. Communities at the same level of integration show evidence of an array of internal governance institutions, displaying "continual local innovation of community governance institutions" (Antinori and Rausser 2010, 11). Each variant in collective action decisions emerged in response to particular problems that particular communities were facing, and these are creative self-organizing responses to local problems.

THE PERMANENT TENSION

The complex governance challenges represented by these institutional choices generate many tensions. Conflicts over employment, decision management, and control are common, leading to a "permanent tension" (López Arzola and Geréz Fernández 1993) between community traditions and CFEs. These tensions are manifested in the (a) hierarchical relationships required in enterprises versus democratic community governance, (b) inefficiencies in the interplay between community needs and norms and enterprise demands, (c) issues of corruption and power disparities, which can led to violence, and (d) conflicts over objectives (Antinori and Bray 2005).

HIERARCHY VERSUS DEMOCRATIC COMMUNITY GOVERNANCE

Community assemblies historically are not skilled in the technical, financial, and management issues of the CFE. In Oaxaca, community assemblies may have more than 300 people, and "many people begin by saying 'no' to whatever proposal is made and do not give proposals serious thought" (Tanaka 2012, 56). Further, community members who are also employees may not appreciate the need for enterprise discipline, a common problem in production cooperatives such as the Basque Mondragon system and Native American enterprises (Jorgensen and Taylor 2000). The tension in Mexico is between the Assembly, individual community members, and the Comisariado. The latter has legal authority but may have difficulty asserting it. As one Comisariado noted, "For example, I'm the Forest Foreman. That gives me authority over you, and I yell at you. Then, you say to me, listen, don't yell at me, this is my enterprise too"

(Gijsbers n.d.). As seen above, the more vertically integrated communities have various ways of delegating authority from the Assembly, allowing managers more leeway to manage (Antinori and Rausser 2003; Bray and Merino-Pérez 2003). In contrast, democratic governance is limited, since commonly only a few women are legal community members (normally widows) and are not allowed to vote in General Assemblies. In the more vertically integrated communities, the problem of hierarchy-versus-democratic governance has tended to smooth out over time with a generally higher level of knowledge of the demands of industrial forestry and the delegation of authority to specialized governance bodies and more professionalized managers. However, As H. Tanaka (2012, 60) correctly notes, "There are good reasons for CFEs to maintain their democratic leadership model. However, this model is not necessarily optimal for the pursuit of excellence in business."

TENSIONS BETWEEN ENTERPRISE EFFICIENCY AND COMMUNITY DEMANDS

Community governance positions must be reelected every three years, though performance issues can lead to shorter terms and some communities change CFE managers with frequency (PROCYMAF 2000). Shorter terms are often preferred by communities as a measure against corruption (see "Turbulence and Violence in Mexican CFEs" below for further coverage of corruption) and to equitably distribute employment, but the CFE then incurs costs when frequently skilled, experienced community members leave and community members with lower skills assume responsibilities. The processes of social learning in this realm can be quite slow. In Oaxaca, it took one large vertically integrated CFE seven years to end the manager rotation system and another one seventeen years. A third one still resists and rotates managers frequently despite high costs to the CFE. Planned rotation in managerial posts occurs with unplanned rotation on the shop floor. Some community members still farm and have other employment opportunities, so may not stay with the CFE for a long period of time. In one community furniture factory with thirty-five to forty employees, twenty to twenty-five workers enter and leave every year, and there is a change of managers every 1.5 years on average. Some community members also do not appreciate that they are also owners, and a more disciplined work ethic could be to their benefit. As Tanaka notes, the enterprises "disrupt embedded norms in one way or another. Therefore, social entrepreneurship does not necessarily enjoy cognitive legitimacy in local society" (2012, 67). CFEs also suffer from pressures to pay high wages and good benefits, which can impact prices and profits. In most social enterprises distribution of profits to shareholders is not permitted, but community members as owners can receive them (Tanaka 2012). Social enterprises also do not have the luxury of simply hiring the most qualified candidate. Hiring from the community, "even if they have suboptimal skills, is part of the mandate of a social enterprise, and it is not easy to fire an employee who has a low level of productivity" (2012, 52).

MULTIPLE OBJECTIVES

The primary objective of most CFEs is employment generation, not profits, but this goal may conflict with maintaining viability in the marketplace. For example, there are tensions between profits and wage policy. Some communities adopt the policy of paying by volume produced to encourage productivity, while others pay a daily wage, which can create a disincentive to productivity (Alatorre Frenk 2000). The community of San Pedro el Alto is aware that it should invest more in its water-bottling enterprise (operated by women) but has not done so since it would reduce the profit distribution. Voting members of the community are individuals with differences in age, income level, experience, migration history, and gender and therefore may have different objectives for those reasons. There may be conflicts between different uses of the forest. There are commonly more conservation-oriented members in communities who oppose logging as detrimental to the forest (Bray and Merino-Pérez 2002). In the Sierra Tarahumara of Chihuahua it is common in many communities for some members to use the forest for grazing livestock, which inhibits forest recovery after logging (Bray and Duran-Medina 2014).

TURBULENCE AND VIOLENCE IN MEXICAN CFES

In likely the majority of cases, Mexican forest communities are able to muddle through myriad internal problems connected with the reality of an entire community trying to operate a business on the basis of a communally owned property, despite this scenario sounding like a recipe for dysfunction. I will classify these management problems into two broad categories—"normal turbulence" and "dysfunctional turbulence"—the latter commonly as a result of corruption and/or organized crime. Turbulence has been defined as actors performing in "uncoordinated and dissonant ways in attempting to meet their individual objectives, typically externalizing as many of the costs and internalizing as many of the benefits of their actions as they can" (Carley and Christie 1992, 156). The notion of turbulence is related to the idea of power conflict in cross-scale networks (Adger, Brown, and Tompkins 2005) but can also refer to conflicts internal to communities. "Normal" turbulence refers to internal conflicts and/or external pressures that require constant negotiation and create frequent tensions, but that do not prevent the CFE from continuing to operate and generate benefits for the community, frequently for years on end. Dysfunctional turbulence includes all forms of disorganization, conflict, and corruption that may result in extreme inequities and internal or external repression and may result in the CFE ceasing to operate, or result in homicidal violence within the community. Some degree of turbulence may be inevitable in a community managing an enterprise, though many communities seem able to substantially avoid it. We saw overall harmony in the case of San Pedro el Alto, where extremely strong community norms of cooperation seem to mute conflict and control any rational egoists who may exist in the community. In the Oaxaca community of Ixtlán de Juárez, Mathews (2011, 231) found that "community trust in forest management is produced by relatively equal contending factions, rather than being imposed by one group alone" and that widespread knowledge of forest management in the community allows members to keep their own professionals accountable. In a case study of a six-community organization focused on conservation in the Chinantla of Oaxaca (see chapter 7 for more detail), conflicts were resolved in both local and national arenas, such as an international conservation congress, through intense discussions and negotiations in contexts of frequent turbulence. It is a measure of the communities' resilience that the process has continued moving forward despite the turbulence (Bray, Durán, and Molina-González 2012).

Other communities may have conflicts that debilitate and destroy their CFEs, up to and including homicides of unclear origin, as we saw in the case study of El Balcón. Activists and academics close to CFEs can become frustrated because some communities can seem to be in states of permanent turmoil and internal conflict over the management of the CFEs. As E. M. Szekely and S. Madrid (1990, 390) have noted, "It is important to recognize some elements of community life as fundamental: the oft-cited consensual decision-making, ritual offices, voluntary community labor, feast days, and other moments in community life. However, we feel that consensus does not imply uniformity but the contrary, the harmonization of different interests, with a focus on the common good." Thus, the "harmonization of different interests" is an endless process in democratic societies and perhaps especially in small communities.

Corruption is a major cause of turbulence and can be considered a form of elite capture. Evidence of community-level corruption surrounding forestry issues is difficult to obtain. Nonetheless, corruption and theft are persistent problems, leading to a variety of outcomes. In the Guerrero community of Cordon Grande, a community leader stole the logging profits for the year and disappeared. Nonetheless, community resilience was demonstrated when logging was resumed in the following year with no associated violence (Durán et al. 2011). Elite capture of government funds for sawmills and other government subsidies or by intercommunity organizations have been reported for Quintana Roo (E. A. Ellis et al. 2015). Another widespread form of corruption, which may lead to a somewhat better distribution of resources, is the granting of loans to ejidatarios that are never repaid. For example, the previously mentioned ejido Pueblo Nuevo in Durango in 1989 had 1,498 ejidatarios, only 90 of whom did *not* request and receive "loans" of widely varying amounts. Ejido La Victoria also in Durango, prohibited loans to individuals, showing that clear and enforced rules can ameliorate these situations (Guerra Lizárraga 1991). This kind of problem can wax and wane. In February 2019 I visited a CFE in Durango where the Comisariado told me that there had been a serious problem with unpaid loans to community members in the previous community government but that he was now assuring repayment by deducting the owed money from profit sharing. In the 1990s, in El Largo in Chihuahua—the largest forest ejido in Mexico, with 266,111 hectares and 1,600 ejidatarios—a political elite dominated profits from the CFE and was accused of corruption. The members who own logging trucks formed a group of hegemonic power within the ejido, with privileged relations with the officials of the Secretary of Agrarian Reform, which has helped them maintain their positions as community authorities and in the CFE, while poorer members of the community are paid low wages (López Álvarez 1994, 29).

Luis Vázquez León (1992) provides a detailed description of the factionalism and corruption that beset the CFE in Santa Cruz Tanaco, Michoacán, which was considered a model in the late 1970s. Internal turbulence in the case of the CFE in Cherán, Michoacán, led to a reduction in logging volume from 42,000 cubic meters in 1985 to 10,000 in 1991, and it was besieged by demands from community owners of carpentry shops for below-cost sales.

Communities also commonly feel that they do not have the power to oppose corrupt leadership. In Cofre de Perote in the early 1990s, for example, the community decided to spend money on two classrooms, but when a cacique spent it on the fiesta, nobody protested (Aguilar et al. 1990). The magnitude and relationship of illegal logging and organized crime were discussed in chapter 3, but in this context there are also frequent reports of violence associated with illegal logging, though normally in communities that don't have CFEs. In the state of Mexico the son of a community leader who attempted to stop illegal logging in the non-CFE community of San Juan Atzingo was killed, and he himself had been jailed on what have been called invented charges (Ureste 2016). According to reports, extensive illegal logging by organized groups is impacting Tarahumara indigenous peoples in the region of Creel, Chihuahua (Breach Velducea 2016), and community leaders in the Chihuahua CFE community of La Trinidad have been assassinated for unclear motives. An uprising by citizens of Cherán against violent illegal loggers in their forests received international attention (Boyer 2015; Zabludovsky 2012), but as of 2019 the community has regained control of its forests (personal communication, S. Anta, March 18, 2019).

Severe problems with corruption and elite capture have also been documented in the pseudonymous community of San Martín Ocotlán, in Oaxaca. In this case, communal institutions failed to check the problems, and a good portrait is provided of how this is done:

> The forestry elite dominates communal institutions through intimidation, manipulating elections, dodging oversight, and discouraging participation in community assemblies. Threats, violence, bribes, and the manipulation of reciprocal obligations are common tools of internal politics. "Some threaten, others invite you to drink," is the way one community member put it. Another vocal dissident of the forestry elite says he fears for his life, and one left town to avoid problems. Dissidents demanding audits of the forestry business complain that the authorities who administered past forestry businesses have taken them aside and told them to desist. A clique of feared leaders uses these pressures to extend their influence over elected community authorities who are not in their camp. The elite comprise the majority on the Council of Distinguished Men, a traditional body of authority parallel to the general assembly, and this traditional institution provides a convenient lever of power for the forestry elite, at times circumventing the community assembly in decision-making. (Klooster 2000, 7)

San Martin Ocotlán, an unusual case for Oaxaca in that the community is ethnically divided, has a community center dominated by mestizos and the outlying communities dominated by Mixtecs.

Although rare in Oaxaca, ethnic divisions with mestizos' dominance of the benefits from the CFE are common in Chihuahua, shown by the marginalization of the Rarámuri and other indigenous peoples and by covert privatization. Some Rarámuri may still have little grasp of the ejido system or are actively intimidated from participating in ejido meetings (Gingrich 1993; Lartigue 1983). This imbalance—which has led to a circle of corruption, comprising caciques within the ejido and logging companies allied with forestry engineers and public officials—includes more than 400 unresolved complaints of illegal logging in a four-year period in Chihuahua, though this level of illegal logging was said to be under greater control later in the 2000s (Pérez-Cirera 2004). Ethnic differences in Chihuahua display power inequalities in common property resource management that impact collective benefits and costs of common property forest governance (Pérez-Cirera and Lovett 2006). In a sample of thirty-eight communities, researchers found that more powerful mestizo authorities have less incentive to cooperate with the entire community and that greater power inequalities lead to higher rates of illegal logging and forest degradation and more inequitable distributions of income. The degree of economic inequality is striking; 80 percent of the private assets related to timber extraction (such as logging trucks) are held by less than 5 percent of the members and mestizos dominate employment in the CFEs (Pérez-Cirera 2004). In some cases the Rarámuri have fought this political and economic domination. For example, in Chinatú in Chihuahua, which is some 80 percent indigenous, mestizos controlled logging until the early 1990s, when work groups emerged as a response to the mestizo domination (Gingrich 1993). In the 1990s, indigenous peoples in the ejidos of Cusárare, Montverde, and Ocóviachi called for audits and conducted sit-ins to protest what they consider to be corruption in the management of logging (Guerrero, Reed, and Vegter 2000). Conflicts between Rarámuri and mestizos within the same community have also exploded into violence in recent years. The Rarámuri leader of the community of Coloradas de la Virgen in the famously violent Chihuahua municipality of Guadalupe y Calvo was assassinated in October 2018 in a fight to get a logging permit annulled that had been issued to the mestizos in the community (Estrada, Villalpando, and Olivares 2018).

VIOLENCE AND ORGANIZED CRIME

Lethal violence associated with forests can be associated with organized crime, guerillas, and state violence, all instances in which the state has not been able to provide an institutional framework that assures CFEs can operate without the threat of violence and organized crime. For example, a 1995 massacre at Aguas Blancas (referred to in chapter 4) by Guerrero state police of seventeen peasants who were protesting illegal logging in their community led to the birth of a new guerilla movement. In 1996, also in Guerrero and also mentioned in chapter 4, a grassroots movement known as the "ecologist peasants" emerged in the Coyuquilla watershed in Petatlán municipality, protesting perceived high levels of deforestation caused by logging controlled by local caciques and the timber transnational Boise Cascade. The movement, sparked by the short-lived incursion of the foreign timber company into local logging operations, subsequently became immersed in forest-related killings and the culture of rural violence in Guerrero, with most reported victims among the ecologist peasants but also with victims among local leaders opposed to them, soldiers, and others.

Organized crime associated with opium poppy and marijuana cultivation, which also branches into other forms of organized crime, is likely the most serious problem facing Mexican CFEs currently. We have seen how the community of El Balcón, which had appeared to be peaceful for decades, has become heavily impacted by lethal violence likely connected to organized crime in Guerrero. Relatively peaceful and successful ejidos have also been impacted by organized crime; FSC certification evaluators for one ejido in Durango were told that under no circumstances should they be on the road at night. CFEs in Michoacán and Durango are reported to have to pay protection money to local crime groups. Multiple reports in Durango and Chihuahua of CFEs discuss how organized crime has taken over the CFE in order to siphon off profits, including in some of the largest CFEs in the region. There are fears that organized crime may be penetrating some large community entrepreneurial alliances in northern Mexico. In Chihuahua, community leaders have been assassinated, apparently linked to organized crime. CFE communities in Durango suffered severe violence particularly in 2010–11, when two cartels battled over influence in the sierra, and currently coexist with the presence of the victorious cartel in a precarious peace (personal observation, February 2019). In general, however, the situation in Chihuahua is considered more violent and dangerous than in

Durango. As of the late 2010s there are municipalities in the Sierra of Chihuahua where foresters only fly in, since road travel is considered too dangerous. It is unknown how many CFEs are exposed to various magnitudes of impacts of organized crime, but since Durango and Chihuahua have the highest number of CFEs, it is very possibly a significant percentage of the total. By contrast, in the state of Oaxaca, the impact of organized crime appears to be quite minor.

In this chapter, I have shown how the basic template for community governance and the harvest rules in CFEs came from above in Mexican agrarian, forest, and environmental laws. These rules from above have not crowded out collective action around forest management. Instead, unburdened from the transaction costs of creating these rules, community collective action has focused on institutions and organization of the CFE market enterprises based on the forest common property. The organizational models were in some cases provided from above through state programs, but the communities took these models and made them their own, finding optimal solutions to local problems. The permanent tension of a community administering a market-oriented enterprise leads to multiple problems, but most CFEs have been able to manage them over decades, facilitated by the historical evolution of a multilevel governance process. However, CFEs also encounter significant internal turbulence through corruption and external pressures, especially from organized crime. Particularly in the latter cases, the larger institutional framework of Mexican government fails to provide the basic guarantees of law and order that allow citizens and businesses, whether community or otherwise, to perform in the market free of fear of criminal elements. The threat from uncontrolled violence and organized crime in several states represents the most serious contemporary threat to the CFE sector.

CHAPTER 6

MARKETS AND THE ECONOMICS OF MEXICAN COMMUNITY FOREST ENTERPRISES

With Juan Manuel Torres-Rojo

T he Mexican community forestry SES emerged from a historic process conditioned by the political economy of forest reforms in Mexico. A space opened that allowed communities to develop their own enterprises based on an economically valuable resource over which they had full ownership. The resilience of the sector depends not only on social capital or social networks, as commonly proposed, but on the community's ability to survive in the marketplace. The conceptual model in chapter 1 shows that markets, along with state policy, is a major driver of the SES and, in interaction with community response, created the framework for collective action and the institutions and organizations that constitute the sector. CFEs are not primarily driven by the profit motive. However, profits and the price mechanism are a major motivation for collective action around forests since they greatly facilitate the primary business goal of employment generation and community public goods. In the preceding chapters, we have seen how Mexican CFEs have emerged in a dialectic between frequently supportive government policy and community agency. In chapters 2–5 the focus has been on the historical evolution of the framework of rules and rights as constituted by the Mexican government. This framework created the space for the vigorous entrepreneurial responses of communities of subsistence corn farmers in defense of their interests and responding to the institutional incentives supplied by the framework.

This chapter focuses on CFEs as economic entities and actors, responding to incentives from both public policy and markets. It examines what is unique about Mexican CFEs as firms and the price incentives that drive the strong collective action around enterprise organization on the forest commons. It then examines the issue of CFE profitability: do they turn a profit, and, if so, how are these profits channeled into the employment and other objectives of CFEs? The chapter also looks at the impact of ongoing forest programs for poverty alleviation and examines the vertical integration strategies of the firms and collective economic action by the firms at the regional, state, and national levels through forest associations. Two particularly noteworthy cases of intercommunity entrepreneurial alliances are presented, and the national and international positioning of CFEs in highly competitive market environments is analyzed.

Mexican CFEs have unusual characteristics as firms (Antinori and Bray 2005). However, like most firms, profits and price incentives play a key role but, unlike most firms, profit maximization does not. Price incentives that allow for profits are normal in firms, but the role of prices as a stimulus for collective action around the commons is ambiguous. Ostrom notes that her examination of CPRs assumes no possibility of forming cartels and "no power in final-goods markets" (1990, 31) and that "it is . . . not a judicious theoretical strategy to presume that choices about rules are made to maximize some single observable strategy" such as profits (207). It is true that CFEs do not seek to maximize profits, but some Mexican CFEs have had power in final-goods market, as we have seen, and have occasionally exercised that power as price cartels in Quintana Roo and Durango and elsewhere (Mota-Villanueva 2002; Vega and Keenan 2014). Mexican CFEs do not pursue a single strategy and clearly have multiple objectives (Antinori and Bray 2005). All of this is to say that Mexican CFEs have many unique features as firms, the issue to which we will now turn.

MEXICAN CFES AS SOCIAL ENTERPRISES AND UNIQUE FIRMS

Relatively high and stable timber prices, despite a decline in real prices in recent years, have sent the signal to CFEs that collective action around forest management will consistently yield greater benefits than costs, particularly the substantial transaction costs associated with maintaining CFE institutions and organizations. With their social mission and lack of interest in profit maximization, CFEs can be seen as an expression of a particular business type, a "social enterprise" (Tanaka 2012) or a kind of grassroots "B corporation," where profits and purpose are balanced as goals (Kim et al. 2017). Social enterprises are usually defined as profit-oriented enterprises that exist to create jobs for disadvantaged populations, such as the homeless or former drug addicts, or to channel profits to social causes. However, social enterprises commonly have tensions between the disparate goals of profit making, surviving in competitive markets, and serving the target group or social cause. In the case of a Mexican CFE, the target population and the social cause are both the community, which was historically disadvantaged, with almost all having their origins as poor subsistence-farming communities. Tanaka proposes that in Mexican CFEs "the communal authority functions like a parent nonprofit" where "the communal authority offers a variety of social service programs that invest the revenues from the community enterprises" (2012, 50). Similarly, the Assembly can be seen as "shareholder's meetings" (Antinori 2005) and organizational forms such as the Council of Principals in El Balcón can be seen as a board of directors (Bray and Merino-Pérez 2003).

Existing theory can be applied only up to a certain point because there are few social enterprises in the literature that are owned by communities operating a common property resource. What makes CFEs unique social enterprises also makes them unique firms in general. As an illustration of this, Antinori and Bray (2005) take a formal institutional economics perspective, suggesting that Mexican CFEs occupy a distinct niche as a form of enterprise organization.[6] Mexican CFEs have a particular pattern of rights with respect to ownership and control, strongly determined by the Mexican Constitution and agrarian law at the constitutional-choice level. A CFE is governed within a formal legal institutional framework that regulates, to varying degrees, both internal and external relations. They are also enterprises that merge the roles of owners, managers, and workers. With respect to internal relations, the theory of the firm is relevant, since it addresses the internal dynamics of production organizations. As noted in chapter 1, Ostrom (1990) explicitly critiqued the theory of the firm as not providing an explanation for how groups of individuals can solve the collective action problem outside of firms or the state. However, Mexican CFEs are formed by defined-member communities, not just groups of individuals, and solve the collective action dilemma by organizing firms in competitive markets based on the common property forest, a situation not considered by Ostrom.

6. The following discussion is adapted from Antinori and Bray (2005).

The emergence of a firm from community governance also creates tensions not normally found in businesses. These include tensions around democracy versus hierarchy, managerial efficiency versus representativeness and tradition, and management for conservation versus timber production, and these tensions have generated a wide variety of enterprise governance forms, as covered in chapter 5.

Communities have been little analyzed as economic actors. Novel governance dilemmas are introduced by collective appropriation of the commons for markets as opposed to individual appropriation for subsistence or minor market sales, the latter the most commonly analyzed cases. A community of subsistence corn farms organizing itself as market-oriented enterprise, a firm, emerges in the dialect between state policy, markets, and community agency, as we have seen. But what kind of firm is a CFE exactly? R. H. Coase (1937) and O. E. Williamson (1975) regard firms as hierarchical command relationships between managers and workers that reduce the transaction costs of constant negotiation, monitoring, and enforcement of contracts. Firms are defined as a "nexus of contracts" between individuals, not a hierarchical authority, where managers receive residual profits as salary, giving them an incentive to monitor workers (Alchian and Demsetz 1972). Firms can have tensions around decision-making with ownership by "principals," who assume the risks of management decisions, separated from "agents," who make daily decisions. The costs incurred by the principals are offset by the benefits of contracting with specialized managers and workers. Whether the emphasis is on ownership or control issues, institutional theories of the firm are studies in collective action regulated by rules (O. E. Williamson 1975). Table 3 reviews various forms of firms as collective action and compares them to CFEs.

The first column presents central characteristics of production organization by separation of ownership and control (Fama and Jensen 1983). Owners receive residual profits after deducting all costs and debts but also assume risks on decisions made. Decision management refers to those who propose initiatives and implement decisions, while decision control refers to those who approve and supervise decisions. Formal legal institutional frameworks and firm objectives have also been identified as key in the literature. Multiple variations in each enterprise type mean that this table is only a sketch of the possible variations. The second to sixth columns present distinct enterprise forms that could carry out forest management operations. Beginning with column 2, private individuals operating NIPFs or "family forests" own small forests with no wood-processing facilities and may contract with outside loggers annually

TABLE 3. OWNERSHIP AND CONTROL IN PRODUCTION ORGANIZATIONS

INSTITUTIONAL COMPONENT	NON-INDUSTRIAL PRIVATE FORESTS (NIPFS)	CONVENTIONAL FIRM	COOPERATIVES		MEXICAN CFE
			INDUSTRIAL	AGRICULTURAL	
Owner(s)	Individual or organization	Shareholders, investors	Labor	Land held by public, community, or individual with sales to farmer-owned enterprise	Official members of the community
Decision management	Owner	Managers	Management committee elected by workers	Management committee elected by producers	Comisariado elected by community members
Decision control	Owner	Executive officers, shareholders, auditors	General Assembly of workers, auditors	General Assembly of producers, auditors	Assembly
Legal system	Land use, tax, environmental laws	Land use, corporate and tax law	Land use, corporate and tax law	National and state cooperative laws	Agrarian, forest, and environmental law
Objectives, assumed or stated	Profit, amenities, NTFPs, bequest	Profit, return on investment	Dividends per worker	Unit price, producer and consumer surplus, services to members	Employment, public goods and services, bequest, amenities, profit

Source: Modified from Antinori and Bray (2005).

or less frequently. The individual owner assumes risk; controls and manages decisions on harvesting, reforestation, and stand improvement; and displays multiple objectives, from profit to conservation. The third column presents the conventional investor-owned firm. A variety of options for compensation for managers and employees exist, but in most assumption of risk and decision of management, on the one hand, and control, on the other hand, are separate.

For publicly traded companies, stockholders permit the board of directors and managers to control and implement decisions, shares are clearly defined and tradeable, and the price mechanism is a check assuring that managers pursue profit maximization as the primary objective.

Industrial cooperatives (fourth column) confine ownership rights to worker members with physical capital assets in common (Jones and Svejnar 1982). Issues similar to the collective decision-making of CFEs arise in the process of electing management committees and a board of directors. However, the worker-owned firm, dissimilar to a CFE, has defined ownership shares so that members can normally redeem shares if they leave the cooperative after a specified time period. The objectives of worker-owned firms are normally the maximization of the dividends to workers. Agricultural cooperatives are normally based on production on individual farms; their collective action takes place in marketing, purchasing, services, or negotiations with external agents and may have multiple objectives. Decision control and management practices are similar to industrial cooperatives. Many of these features are shared by CFEs, but managing a common property territory distinguishes them from agricultural cooperatives.

The last column notes some of the unique characteristics of CFEs as firms. Only official registered community members, normally by birth and upon reaching a certain age and fulfilling community service requirements, can be owners. CFEs do not have defined shares that are tradeable in the marketplace. Risk is spread among all community members equally, but members who opt to leave the community may give up all rights to profits and products from the forest, frequently a deterrent to exit. Community members are full owners of the forested territory, natural capital, after the 1992 constitutional amendment, and the CFE is based on the common property forest. All of these characteristics distinguish CFEs from the other firm types. With respect to decision management and control, there are some similarities with the other enterprise forms, where assemblies of stockholders make policy decisions. In both cooperatives and CFEs, there can be conflict between community members as owners and workers. Decisions over benefit flows are especially important. In many cases in Mexico, much of the CFE benefits are distributed in public goods. Finally, all forms of enterprise exist in a regulatory legal environment, and in Mexico agrarian, forest, and environmental laws play a significant role in shaping CFEs, as we have seen in earlier chapters. To summarize, a CFE can be defined as a firm based on collective ownership of a forest by a community, with governance derived from agrarian law but infused by community rule-making around the specific enterprise form but with very limited influence over harvest rules.

CFEs commonly experience tensions between direct "democratic" community control and hierarchical management structures and typically have multiple goals with profits serving as an intermediate objective for the ultimate objectives of employment, public goods, and household income. Thus, "the community as entrepreneurial firm" (Antinori 2005) can be seen as a variant of private corporate production or as a "third way" between private- and public-sector production. However, as a kind of firm, the community cannot be placed in Ostrom's conception of collective action around the commons as a "third way" between firms and the public sector, as noted in chapter 1. Neither the conception of CFEs as social enterprises nor as unique firms should obscure the also unique and remarkable historical transformation that has taken place. CFEs originate in agrarian communities of individual subsistence corn farmers that have transformed themselves into collective action organizations around market-oriented common property forest enterprises, sometimes with far-reaching consequences for community social organization. As has been noted, CFEs "have altered the social relations of the community of origin and, in a complex political process of social reengineering, have constructed a new social order . . . [I]n contrast with the traditional campesino societies where families are the organizing center of economic reproduction . . . in these communities that axis appears to move towards a collectivist industrial economy of village reach and towards a social order of a community-corporatist character" (Garibay Orozco 2008, 435). Garibay Orozco's characterization likely applies only to some of the largest and most successful CFEs, but for those it effectively represents the magnitude of the transformation that has taken place. However, these remarkable examples of social reengineering also depend on the ability to survive in national and international markets to assure the achievement of the social goals. This situation brings us to the question, How has the sector been able to survive and remain relatively resilient in the face of national and international competition some fifty years since it began to emerge?

NATIONAL AND INTERNATIONAL COMPETITIVENESS AND THE PRICE HISTORY OF PINE AND TROPICAL TIMBER (MAHOGANY AND CEDAR)

Mexico appears to have few competitive advantages in maintaining a national market for its timber in the face of international competition or in competing in international markets. The United States has a forest area that is four times the size of Mexico's but that carries out fifty times the production, and Chile has a forest area that is one-quarter the size of Mexico's but carries out three times the production (Forster et al. 2004; Kaimowitz 2004). Mexican production forests and sawmills are on average extremely small compared to international standards. Ninety percent of sawnwood production is pine and is not competitive in export markets. Production of tropical hardwoods has declined because it is no longer used in railroad ties since cheaper cement substitutes are now commonly employed. Historically mahogany (*Swietenia macrophylla* King) and Spanish cedar (*Cedrela odorata L.*) for fine furniture have been important, but production of these species has been in decline due to historical overharvesting (see chapter 7). A variety of the so-called lesser-known tropical species dominate the market, but they must compete with imports from Brazil and elsewhere. CFEs dominate roundwood production, with some 85 percent of the national production, though only around 15 percent of sawnwood production comes from community sawmills (G. Chapela y Mendoza 2012).

In addition to the threats from imports, many CFEs have myriad deficiencies as businesses in a competitive marketplace, as noted in chapter 1. These weaknesses may include harvests below the annual forest growth rate, decapitalization, low investment in technology and modernization, small production forests, inefficient sawmills, poor-quality classification practices, managerial and organizational problems, high transportation costs and other issues, many of them inherent in communities trying to administer a market-oriented enterprise (G. Chapela y Mendoza 2012). As well, a 2002 law established taxation of 30 percent of CFE profits for sawnwood; this tax significantly reduces profits and occurs despite the fact that much profit is used to provide public works that should be supplied by federal tax dollars (S. Anta, personal communication, April 19, 2018).

Due to these issues and others, Mexico has run persistent trade deficits in forest products for decades; national production only covers some 20–30 percent of consumption. Mexico consumes around 35 million cubic meters of wood annually, but national production is less than 7 million cubic meters. For example, some 90 percent of plywood consumption is imported. However, it is crucial to note that Mexican sawnwood is more competitive nationally—only around 50 percent is imported. Sawnwood prices are determined by a series of criteria that include board dimension and ring growth position, dependable quality classification, kiln drying, edging and planing, and wood appearance. Sales and service are also important, incorporating issues such as timeliness of delivery, quantities of desired product, and responsiveness to buyer needs and complaints.

Imported pinewood comes mainly from Chile and the United States, followed by Canada, Brazil, and Venezuela. Chilean timber is considered inferior to Mexican timber in its aesthetic and technological properties but superior in price, presentation, and quality of service (Forster et al. 2004; Kaimowitz 2004). Despite all these challenges, the majority of CFEs, as we have seen, continue to operate year after year for decades, even if some of them choose to reduce their level of vertical integration. So how are they surviving?

To answer this question we will first look at the price history (adjusted for inflation and exchange rate) for recent decades of pine and tropical timber. Figure 13 shows the export prices for pine and all tropical timber sawnwood (*escuadría*) over a forty-six-year period (1970–2016).

Pine is by far the most important tree species from temperate forests. Tropical timber was historically dominated by mahogany; however, in recent decades its share of total production has fluctuated from between 15 percent and 27 percent for Quintana Roo, the principal producing state. Tropical timber includes the harvest of nineteen lesser-known tropical species and thirty species of polewood (E. A. Ellis et al. 2015), though these export prices reflect mostly mahogany, since few other tropical timbers are exported. Figure 13 shows that prices for pine were relatively stable from 1970 to 1978 in the US$300 per cubic meter range, with sharper fluctuations between US$200 and US$600 between

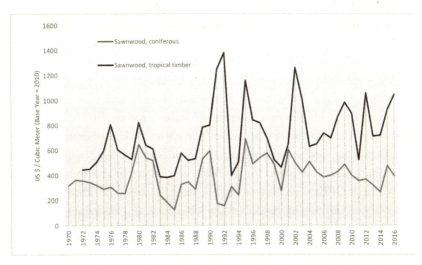

FIGURE 13. Prices for pine and tropical timber sawnwood (1970–2016). Graph courtesy of FAOSTAT (n.d.).

1978 and 1994. However, after 1994 fluctuations were dampened, commanding relatively high and relatively stable prices, averaging around US$400 per cubic meter since 2002. Tropical timber prices have fluctuated much more sharply, though they have stabilized a bit more since around 2002. B. Hodgdon (personal communication, April 12, 2019) argues that the position of tropical timber CFEs is particularly perilous due to competition from plantation mahogany in the Pacific islands. He argues that lack of diversification, decreasing productivity of mahogany, and underharvesting of lesser-known species (LKS) puts at risk the future of tropical CFEs.

Part of answering the question of CFEs' survival involves understanding how these prices compare with the international competition and the degree to which price has served as an incentive for collective action, focusing on pine wood since it is by far more economically important. National timber production is under pressure from lower-cost imports due to multiple trade agreements for wood and wood products that have created open frontiers. Under NAFTA, timber and wood products can be freely imported from the United States and Canada with no tariffs (and the same is true for Mexican wood exports to those two countries). Since 2006 Chilean timber is also free of all tariffs (G. Chapela y Mendoza 2012). Mexico imports wood products mainly from the United States and Chile but imports from Brazil, Peru, Argentina, and Uruguay have also been increasing. For example, between 2010 and 2015 imports of Brazilian wood products to Mexico increased over sevenfold, from US$18 million in 2010 to US$139 million in 2015 (ITTO 2016). To understand the impact of this international competition, Figure 14 shows the pine sawnwood price comparison for Mexico, Chile, and the United States from 1970 to 2016.

The graph shows there have been sharp variations in the price of Mexican pine sawnwood, most commonly due to inflation and exchange rate fluctuations during periods of economic crisis in Mexico. Historically, Mexican prices have been generally higher than prices from either the United States or Chile, though since around 2006 Mexican and U.S. prices have been generally similar. The somewhat higher prices for the Mexican product are due to higher costs for Mexican CFEs through the chain of vertical integration. The deficiencies noted above all drive up costs, but low labor costs and labor-intensive practices help compensate for these higher costs. Costs are also higher due to the slower growth of natural forests than of plantations. However, and most important, despite the higher prices and costs, the Mexican market shows a marked preference for the national product due to the higher quality of timber

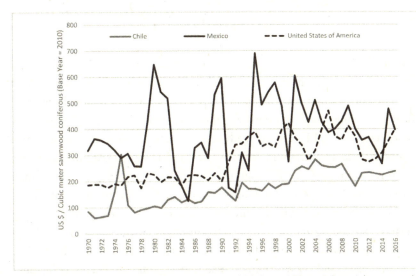

FIGURE 14. Prices for pine sawnwood in Mexico, Chile, and United States (1970–2016). Graph courtesy of FAOSTAT (n.d.).

from native forests. As discussed further in the following section "the large national demand for forest products and solid wood products has allowed the CFEs to have positive profits despite high average production costs" (Cubbage et al. 2015, 646).

The long-rotation national pine has qualities not present in short-rotation plantation pine and is sought after by the construction industry, carpentry shops, and other finished wood producers (Molnar et al. 2008). Mexican timber is dominant in construction, since imported wood only accounts for 10 percent of this market, given the poorer quality of Chilean timber. For warehouse pallets, as another example, Chilean wood only allows for one to two uses, while Mexican timber allows for three or more (G. Chapela y Mendoza 2012). Thus, Mexican community timber is concentrated in high-value market niches in the domestic market (Kaimowitz 2004). These findings are important and highly relevant, and they also point to the need for ongoing government support in human, physical, and financial capital. The continued survival of the majority of Mexican CFEs in the face of lower-priced imports and with few competitive possibilities internationally suggests a high degree of economic resilience, buttressed by their profitability and supported by relatively high and stable prices.

CFE PROFITABILITY

Mexican CFEs have often been claimed to be unprofitable for reasons noted above (G. Chapela y Mendoza 2012). Their flexibility as enterprises is reduced because they cannot shift labor and capital into more profitable businesses, having no other viable entrepreneurial options than the forest (Aguilar et al. 1990). However, CFEs also have advantages. Natural forest capital commonly maintains and appreciates in value, unlike physical capital (Antinori and Bray 2005; Klemperer 1996). CFEs can therefore make the decision to not use their natural capital for a period for multiple reasons, make adjustments in their levels of vertical integration, or restructure the enterprise (as in work groups) to reduce costs or increase profits. Governance institutions in CFEs can result in cost savings, since many use the unpaid cargo or committee systems for administration, and in Oaxaca and elsewhere traditional unpaid community labor obligations (*tequios* or *faenas*) are used for forest and road maintenance (Aguilar et al. 1990; Antinori and Bray 2005). In Oaxaca, almost all communities in a forty-two-community sample held tequios throughout the year, for an average of three days each and from one to five times a year. The least-integrated groups had the most unpaid tequios, lowering their costs, while more integrated communities relied more on wage labor (Antinori 2005).

From early periods, there is only anecdotal evidence for the impact of launching CFE operations on profitability and incomes, but it appears to have been substantial. For example, an ejido in the Plan Puebla doubled its annual income when it started managing its forests (Halhead 1984). In the Oaxaca community of San Miguel Peras, when the community began to operate its own CFE in 1979–80, community profits went up 600 percent and wages to workers doubled (Klooster 1997). Reports such as these are plausible, since the best-documented studies suggest that CFEs at all levels of integration are profitable, though the studies vary in how they estimate profits, land opportunity costs, and other issues. In a study of thirty CFEs in Oaxaca it was found that all were profitable, expressed as gross profit as percent of sales revenue, for all levels of vertical integration and timber products. Type II stumpage communities had 39 percent profit, Type III roundwood communities had 38 percent profit, Type IV sawnwood communities had 54 percent, and Type V finished products communities 32 percent (Antinori 2005). The lower profits for the finished products community are likely due to the higher costs associated with more capital-intensive forms of production. Another study of two Type III and two Type IV communities in Oaxaca found that three of them had very high levels of profitability, one with profits of 106 percent, and one was essentially at a break-even point (Merlet 2015). A study of two large ejidos in Durango, Pueblo Nuevo and San Pablo, found that both were profitable (G. Chapela y Mendoza and Meraz n.d.).

A recent national study of thirty communities (seven Type III communities and twenty-three Type IV communities), evaluated costs, revenues, and profits in three stages of vertical integration: defined as forest management, harvesting, and milling (Cubbage et al. 2015). Ninety percent of the sample were in temperate forests and in these 85 percent of the harvest was pine, 8 percent fir, and 8 percent oak. In the three cases from tropical forests, 15 percent of the harvest was in mahogany and the rest in a wide variety of lesser-known tropical species. This study found that standing inventory, an indicator of potential profits when the inventory is debt free, showed large amounts of mature timber that were more than fifty years old. The average for standing stock was 178 cubic meters / hectare, but six of the communities had standing stock of more than 280 cubic meters / hectare, indicating that they were more than 100 years old, "essentially old growth and very mature timber" (2015, 631). This excess growing stock is one explanation for why many Mexican CFEs remain profitable in the face of foreign competition from plantation timber. "Harvests of relatively rich natural forest estates almost always made forest management for the CFEs profitable" (646). The majority of the communities were also found to be sustainable in terms of their harvesting intensity (measured as the proximity of the harvest rate to mean annual increment), an indicator of profitability in future years. The harvest totaled less than the mean annual increment of the standing stock with positive net present value (NPV) for a thirty-year period, and only two or three of the communities appeared to be unsustainable. The key finding is that "enterprises generally were profitable and sustainable at the same time" (644). Profitability and sustainability of CFEs could become an issue as they liquidate excess growing stock or reduce investment in natural assets in the form of silvicultural practices. However, harvesting and conservation constraints imposed by the FMPs, as well as CFEs' resilience on multiple fronts, reduce the risk of bankruptcy.

Forest management, the first stage of vertical integration in the classification of Cubbage et al. (2015), is defined as activities such as reforestation, periodic management, roads, technical assistance, silviculture, and payment to communities. Harvesting is defined as cutting the timber and hauling it to roadside, while milling includes both hauling timber to the sawmill and processing it into sawnwood. Thus, in this analysis, Type III communities conduct forest management, logging, and hauling to the logging road, and Type IV communities do the first three activities as well as yarding and milling. (It should be noted that this definition does not correspond exactly to the CONAFOR definition of Type III, since CONAFOR's Type III commonly includes hauling logs to the sawmill.) Forest management costs were generally low, since most communities depend on natural regeneration rather than reforestation. All forest products were included, but 90 percent of profits were in timber, with NTFPs generating 7 percent and payment for environmental services 3 percent of profits. Only one of the thirty communities was not profitable at the forest management stage. Harvesting costs (cutting trees and delivering logwood to the roadside) were considered somewhat high but reasonable, and twenty-two of the thirty CFEs made a profit on harvesting. For milling, there was wide variation, but the average sawmilling cost—including the wood, equipment, and operating costs and indirect labor and energy costs—was US$129/ cubic meter, considered moderately expensive. Some CFEs have divided their logging and milling operations into separate businesses, in order to have a better grip on costs in each set of activities. Eighteen of the twenty-three (78 percent) sawmill operations in the sample had profits; only five lost money. These eighteen operations had average profits of US$53/ cubic meter, which were considered "quite large," and, according to Cubbage and colleagues, "returns on investment for the Mexican mills are still generally positive for the entire value chain, based apparently on the relatively high price for lumber in the Mexican market, where demand has been high" (Cubbage et al. 2015, 643). Return on investment in the twenty-three communities ranged from 3 percent to an astounding 445 percent. The 50 percent share of the domestic demand for sawnwood has apparently achieved a degree of insulation from global competition, due to its higher quality, and it continues to sell at relatively high prices (Cubbage et al. 2013). This insulation is an important indicator of the degree to which CFEs literally profit from a historically emergent safe economic and political operating space.

Profitability varies according to the degree of vertical integration, but higher degrees of integration don't necessarily ensure higher profits, as noted above. Cubbage et al. (2015) showed that the greatest profit in CFEs is concentrated in forest management. Although there are few studies, this finding strongly suggests that most Type II roundwood communities are profitable, since CFEs selling logs are more profitable than sawmill communities, though lack of investment in the forest could lead to deteriorating conditions. Nonetheless, it is high prices for stumpage, which allow communities to generate profits from forest management but impede them from competing in export markets. Variations in profitability are strongly associated with the cost structure of the vertical integration stages. It is likely that the higher profits in the forest management stage are related to lower investments in silviculture (such as limited reforestation and thinning practices) and the excess growing stock. But with respect to costs, "overall timber harvesting costs appeared to be in a reasonable range for typically mountainous conditions found on Mexican CFEs" (Cubbage et al. 2015, 639). For sawnwood, production costs are higher, but "much of the old growth timber in Mexico still has been composed of higher grades, rendering higher production costs less detrimental to competitiveness. In addition, prices within Mexico must have been high enough for the CFEs to continue to sell their products within domestic markets" (643).

CFEs were considered to be "fairly sophisticated" in that they had FMPs (required by law as we have seen) and vertical integration through to sawmills. There is evidence that focused technical assistance can have major impacts on profitability. San Bernardino Milpillas Chico, in southern Durango, went from losses of over US$500,000 in 2005 to a 2008 profit of nearly US$1.8 million due to comprehensive support in improving forest management and processing from Rainforest Alliance (Hernández et al. 2010). This kind of benefit from technical assistance confirms the observation that "continued adaptations and improvements in their operations would be required for CFEs to prosper with global competition, but they have had a good forest resource and have demonstrated reasonable management skills and experience to recognize and make such improvements, with capacity building, the technical assistance from government organizations, and continued financial assistance" (Cubbage et al 2015, 647). Evidence of large-scale state technical assistance's effectiveness for poverty alleviation is presented later in this chapter, in "State Programs and Poverty Alleviation."

DISTRIBUTION OF PROFITS AND MULTIPLE OBJECTIVES

Given that there is convincing evidence that CFEs are profitable, what do they do with the profits? In most private-sector enterprises, profits are distributed to shareholders or reinvested in the enterprise, and these are options for CFEs but because CFEs are social enterprises, profits are invested in multiple channels. As has been noted for Oaxacan CFEs, "[Private enterprises] aspire only to maximize their earnings, while the community enterprises seek . . . the generation of sources of employment, the conservation of the forests, the production of resources for collective benefit, and the maximization of the participation of the comuneros" (Alatorre Frenk 2000, 52). The former president of a Oaxaca intercommunity organization has further noted: "Maybe our community enterprise isn't as efficient as a private enterprise. Well, that's just how it is. Our goal is the development of the community, and not the enrichment of a few" (Gijsbers n.d.).

The multiple objectives of CFEs can be classified into four areas: (1) employment and benefits, (2) public goods for the community, (3) reinvestment in the forest enterprise and other community businesses, and (4) profit sharing to legal members of the community. Each of these will be considered in turn.

EMPLOYMENT AND LABOR ARRANGEMENTS

CFEs can generate jobs for very high percentages of the community labor force, depending on the size of the community. There are large CFEs that generate very large numbers of jobs, such as San Juan Nuevo Parangaricutiro in Michoacán, which provides some 900 full-time jobs, and small CFEs that provide employment to nearly everyone in the community who wants it, such as Rosario Del Xico in Veracruz, which provides twenty-four jobs (Bray and Merino-Pérez 2004; Merino et al. 1997). The Pueblo Nuevo ejido in Durango, with over 73,000 hectares of commercial forest, generates 515 direct jobs, only 10 percent of them held by ejidatarios. The rest of the employment goes to other types of legal residents of the community, such as the children of ejidatarios and other people who live in the community (*avecindados*; G. Chapela y Mendoza and Meraz n.d.). The Tepehuán indigenous community of San Bernardino Milpillas Chico in Durango employs 360 people (Cramaussel 2013; Hernández, Fortín, and Butterfield 2010). At the other extreme, there are Type II stumpage communities where almost no one is employed in logging, since most of the work is done by outside contractors. In a sample of forty-one communities in Michoacán and Durango, Type II stumpage communities have most labor coming from outside the community, while Type III extraction communities employ almost entirely community labor, though the percentage falls somewhat in the sawmill communities (Antinori and Rausser 2010). Forestry commonly represents steady employment income for smaller percentages of the total community labor force, suggesting that many people prefer to pursue other income-generating activities. In a study of forty-two CFE communities in Oaxaca, the percentage of community members receiving income from the CFE on a regular basis was 15 percent in stumpage communities, 19 percent in roundwood and sawnwood communities, and 26 percent in Type V secondary products communities (Antinori and Bray 2005). These figures suggest that, at best, and depending on the size of the community, CFEs may only generate employment for up to 25 percent of community members, including both legal members and legal residents. However, as noted by Tanaka (2012), there can be tensions around decisions made by the social enterprise for profit that have a negative impact on community employment. In Ixtlán de Juárez in Oaxaca, negotiations within the community occurred over modernization of a sawmill that would reduce employment, compensated by the creation of a furniture factory that would generate new employment. In Pueblos Mancomunados, also in Oaxaca, the decision to locate a water-bottling plant and a sawmill in Oaxaca City, the rational business decision, was criticized by community members because it reduced employment opportunities within the community.

Access to regular paid employment within the community, in some cases with benefits, is extremely unusual in rural Mexico and is thus significant even when the income is not large. In Chihuahua, a study of seventeen CFEs at all levels of vertical integration, some having generally large populations and relatively low volumes of timber, found that the only source of paid employment in most of the communities was forestry. Only three of the communities provided employment for as many as 40 percent of the community members, and most were below 10 percent. A calculation of full-time equivalent employment in the CFEs found an average of less than 20 percent full-time equivalent employment (Bray and Duran-Medina 2014). Nonetheless, this time dedication spread across the entire labor force likely represents a few months' income in a very low cash economy, and thirteen of the seventeen CFEs in the study reported that forestry was the most important source of cash income. In more prosperous situations, community labor force participation is also not higher because many people are not interested in the hard, dangerous work of logging or the discipline of factory work, if other opportunities or remittance incomes are available. For example, the 26 percent employment level in Type V communities reported for Oaxaca may imply clear ceilings to the interest of working in the CFE. Most of the more prosperous CFEs also support various fringe benefits and social welfare benefits, such as health coverage, accidental death benefits (distressingly common, given the dangers of logging), and old-age pensions (for all legal community members, not just former employees in the CFE), as in the cases of San Pedro el Alto and El Balcón (Torres Rojo, Guevara-Sanginés, and D. Bray 2005). Health, retirement, and accidental death benefits, however modest, generated by a community enterprise, are extraordinary in rural Mexico. There are apparently no studies of the specific impacts of CFE employment on household incomes or total benefits to communities, or the precise wage levels by rural Mexican norms, a significant research lacuna. Finally, CFEs have a wide variety of rules for labor contracting arrangements, in addition to direct employment. For example, many larger CFEs establish contracts with logging truck drivers and winch operators from the communities who recruit their own workers, usually through family networks. The drivers are usually single individuals, but in the case of the winch operators they are responsible for all of the expenses of their logging team (commonly composed of five to seven workers), reducing direct employment responsibilities from the CFEs.

PUBLIC GOODS AND SERVICES

Despite little quantitative evidence available on what percentages of CFEs invest in public goods and their magnitudes, observational evidence suggests that many communities invest significant funds in public goods, such as community infrastructure and services that are normally the responsibility of government. Communities commonly use profits to construct or restore churches, municipal buildings, public lighting, potable water systems, clinics, and schools, as well as sponsor celebrations for religious and civic holidays. In the Durango/Michoacán study of forty-one communities, most public investments went first to schools, followed by community fiestas and medical services (clinics, medical supplies, and salaries for medical staff). In the forty-two-community Oaxaca survey, almost all of the communities channeled some CFE profits to public goods (Antinori and Bray 2005). In El Balcón 60 percent of a million dollars in profits for one year was invested in community public goods, such as improved housing for the entire community (Bray and Merino-Pérez 2003). In the case studies of San Pedro el Alto and El Balcón, we have seen the range of investments in social welfare and public goods that can take place in the most prosperous of the CFEs. In both the FMPs and in land use zoning practices (see chapter 7), it is common to set aside large forest areas for strict conservation.

This practice is, in essence, the demarcation of "private" conservation areas with the consequent provision of ecosystems services as both local and global public goods (see chapter 7).

FINANCING OF CFES

Traditionally and in many cases today, the financing of logging operations comes from advances from logging contractors. Most smaller CFEs in Quintana Roo and elsewhere forward-sell part or all of their expected production to buyers or brokers at a fixed price. The widespread presence of work groups in Quintana Roo means that the advances go to individuals who hold shares of the anticipated volume, with forward sales often made for years in the future (Hodgdon 2014), the futures market referred to in chapter 5. The other major sources are self-financing from profits and subsidies from the state. Historically, in Durango in particular, private enterprise was an important source of financing for capital investments, as in the financing of sawmills and infrastructure for the community of San Bernardino Milpillas Chico in Durango (Hernández, Fortín, and Butterfield 2010). The reinvestment of profits, for forest and road maintenance, equipment, FMPs, and other things, is more pronounced in the Type IV sawnwood and Type V finished products communities. In the Oaxaca study, most communities made such investments, though all more vertically integrated CFEs did so and at higher rates. Logging can be profitable enough, as we have seen, to allow for self-financing of new physical capital, commonly reported in the Oaxaca study. Nearly all of the trucks, cranes, and sawmills owned in the sample of forty-two were self-financed. In the case of trucks, ownership is relatively evenly split between individual and communal, with the incidence of communal truck ownership increasing with integration level. Individual ownership of cranes occurs in Type II and III communities, but in Type IV and Type V cranes are entirely community owned and almost all purchased with profits (Antinori 2005). It is noteworthy that in most cases physical capital was purchased with community or individual funds and relying little on outside sources. In Durango and Michoacán, communities that are members of forest associations consistently invested greater amounts in production (Antinori and Rausser 2010). The degree of self-financing may vary considerably from state to state and historically. We saw in chapter 2 how FONAFE in the 1970s was the main source of financing for sawmills, and in recent years CONAFOR has also been active directly by subsidizing community sawmills through various forest incentive programs or indirectly by financing loan collaterals for communities.

Self-financing and government subsidies are by far the two most critical instruments for acquiring financial capital. Only 7 of 41 communities in the Michoacán/Durango sample had ever received commercial or development bank credit and only 3 of 42 had in the Oaxaca study (Antinori 2005; Antinori and Rausser 2010). With respect to government subsidies, 18 of 33 CFEs in the Michoacán/Durango study received government subsidies for FMPs for an average of 65 percent of the costs (Antinori and Rausser 2010). B. D. Hodgdon (2014) has identified financing as one of the most significant bottlenecks for achieving greater competitiveness for tropical CFEs in Quintana Roo, noting that less than 1 percent of credits for primary production in Mexico go to forestry.

PROFIT SHARING

Communities may also make the decision to distribute all or a percentage of the profits on a proportional basis to all legal members. Communities that decide to distribute most, if not all, of the profits may do so because of mistrust of authorities, poverty, or little possibility of investing in vertical integration. There is a marked tendency for poorer communities to distribute all or nearly all of the profits, while wealthier communities tend to distribute little or none of the profits, concentrating all investments in the enterprise, prioritizing employment generation and public goods. Amounts of annual profit distribution can be as low as a month's average income or as large as an average annual income for rural areas. A Quintana Roo study of five communities showed an average of 13.5 percent and a high of 30.3 percent of household income coming from profit sharing (Armijo Canto and Robertos Jiménez 1997). In the Durango community of Canelas, which is divided into work groups, each work group establishes independent rules for profit sharing, with varying amounts going to community members, public goods, and administrative costs (Taylor 2005). The Oaxaca study showed that less than half (fifteen of forty-two) of the communities distribute profits. The Type IV and V communities that distribute profits only distributed around 20 percent of that distributed by Type II communities (Antinori and Bray 2005). By contrast, in the Durango/Michoacán study all but three community-level CFEs paid out dividends to community members, and all work groups did, and there was little public goods or capital investments in the CFE (Antinori and Rausser 2010).

STATE PROGRAMS AND POVERTY ALLEVIATION

In chapter 3 it was proposed that the community forest sector had largely come into being by 1988, and, after the indifference of the period of Salinas de Gortari, the national government has shown a consistent policy of support for the sector. Chapter 3 also reviewed the multiple government programs that have presumably helped the sector to mature and consolidate its gains, but few evaluations exist of the actual impacts of these programs. This lacuna was significantly addressed by Torres Rojo, Moreno-Sánchez, and Amador-Callejas (2019) in a study of the impact of state programs on poverty alleviation and forest cover conservation (see Alix-Garcia, Sims, and Yañez-Pagans 2015 for a study of the impact of the PES program on poverty alleviation). The study evaluated the Community Forestry Program (the successor to PROCYMAF) in a quasi-experimental and quantitative approach to compare matched treatment and control communities at the preintervention stage and after five years of program support. The study used a CONAFOR database with more than 20,000 records to identify 5,074 of the communities that met study criteria. Poverty impacts were evaluated by use of the official government Marginality Index and forest cover conservation by using the proxies of rate of change on forest cover and fragmentation. The CONAFOR grants were classified as supporting capacity building in the human, social, economic, and environmental dimension (or human, social, financial, physical, and natural capital in the terms used in this book). With respect to the impact on poverty, the study found that communities that received all four forms of support and only support for social and human capital both had statistically significant reductions in poverty over the control groups during the five-year period. However, the communities that received only support for human and social capital actually had higher rates of poverty reduction than the "all-capacities" form of support, suggesting that targeting support to the two areas would be more cost-effective. On the other hand, neither form of support had any statistically significant impact on forest cover conservation. This lack of causation is interpreted to be due to the fact that communities that are actively managing their forests "are already convinced and committed to the sustainable management of their forests" (Torres Rojo, Moreno-Sánchez, and Amador-Callejas 2019, 117). Nonetheless, the findings on poverty alleviation are some of the first quantitative evidence of the effectiveness of state support channeled to community forestry.

CHOICES REGARDING VERTICAL INTEGRATION

The choices community enterprises make with respect to vertical integration "reveal their perceptions of risk, cost, and benefit of engaging in the wood products market" (Antinori 2005, 242). Thus, upward and downward mobility, or maintaining the status quo, in the vertical integration hierarchy tells us some important things about the way internal organizational issues and external market pressures influence CFE decision-making. In chapter 5, we saw how CFEs choose different institutional and organizational strategies to deal with internal political and distributional problems and issues of transaction costs, while in this section we examine how these issues impact the vertical integration decision.

As Antinori and Rausser note, "Ownership and control over the assets used in production, like chainsaws, cranes, trucks, tractors and sawmills, has huge implications for the transaction costs of production," that is, how they achieve both monetary and nonmonetary benefits (2010, 11). In the case of Oaxaca, we saw that historically many CFEs have advanced in vertical integration with reinvested capital from the CFE and that in different historical periods, government programs have also provided important inputs of physical and human capital that allowed communities to rise in vertical integration. For Oaxaca, Antinori documents a "vertical integration transition" in the 1986–97 period. In 1986, of a 44-community sample, 30 communities were Type II, 9 Type III, and 5 Type IV, with sawmills. By 1997, there were 16 Type IIs, 13 Type IIIs, 8 Type IVs, and 7 Type V finished product communities, a tripling of Types IV and V from 5 to 15. Notably, this transition occurred during a period of little government support, and indeed C. Antinori found that none of the CFEs that had vertically integrated to finished products had received government support (Antinori 2000, 2005). This outcome suggests that although government subsidies have clearly been helpful, integration into markets with a demanded product that can fetch good prices has been the essential factor in growth of the sector.

CFEs can also decide to deintegrate vertically, reducing transaction and other costs and accepting lower profits (García-López 2013). It is apparent that there is an ongoing flux both in total numbers of CFEs and numbers at each level of vertical integration: new CFEs commence operation, CFEs cease to operate, and established CFEs experience upward and downward mobility. As noted, state policy has often been helpful in upward mobility, though its deficiencies likely also explain some of the downward mobility. Beginning with the parastatals FONAFE and the DGDF in the 1970s and through the 1990s and

2000s with PROCYMAF and its policy successors, the federal government has, in varying magnitudes of funding, both helped launch new CFEs and provide the capital inputs that allow them to increase vertical integration. In its first years of PROCYMAF's operation in the late 1990s, a total of 41 percent of the parastatal's support went to communities that had never logged before or had not logged for considerable periods (Type I communities).

Downward mobility or vertical deintegration has been documented particularly for the El Salto region in Durango. A study of four forest associations there found that that fifteen communities had sold their sawmills and returned to selling roundwood (García-López 2013). Inefficient sawmills, high transaction costs, emigration, the increasing formation of work groups who don't sell to the community sawmill, and diversification of productive activities were all suggested as causal variables (interview with G. García-López, April 26, 2017). In the Michoacán/Durango study, a notable deintegration occurred from 1997 to 2007, with the number of sawmill communities in the sample declining from twelve to seven (Antinori and Rausser 2010). Tendencies do vary by region and historical period—in Oaxaca from 1987 to 1997, as we have seen, there was a marked shift upward in vertical integration—and there appear to be no clear determining variables in these varying outcomes, suggesting the variation may result from different degrees of social capital and political relationships in the different regions and varying historical contexts. However, recent studies of vertical integration flux in Chihuahua, Guerrero, and Oaxaca over varying time periods suggest an overall decay in vertical integration (G. Chapela y Mendoza 2018b). In Chihuahua in the 2000–2013 period, only 5 CFEs advanced from Type III to Type IV, while 38 CFEs decayed from Type IV to Type III. In Guerrero over a much shorter time period, 32 communities began CFE operations, going from Type I to Type II, but 55 decayed from Type II to Type I—a net loss of 23 CFEs—and another 9 Type III and IV CFEs went out of business. In these two states, there is likely some link to the extremely difficult security conditions prompted by organized crime. In Oaxaca from 2006 to 2012, 10 new CFEs were launched, but 40 CFEs ceased operations, while 11 advanced in vertical integration and only 3 Type IVs decayed to Type III (G. Chapela y Mendoza 2018b). The numbers that have ceased business or experienced downward mobility are one of the signs that have led some observers to suggest a "crisis" in the community forest sector.

In order to obtain a clearer picture of the flux nationally in vertical integration, one of the authors of this chapter conducted an original analysis for this book of 2,967 cases of communities that received support from CONAFOR from 2004 to 2011 and for whom there was a record of their typology at the beginning and end of the period. The support received was classified in thirteen different categories, from support for the FMPs to training (Torres Rojo and Amador Callejas 2015b). Table 4 is a transition matrix that shows the upward and downward flux and stability in the vertical integration hierarchy for the CONAFOR beneficiaries over that period.

This table shows considerable flux in vertical integration among CFEs, though it is also notable that most CFEs remained in the same type at the end of the period. Of the 2,136 Type I at the beginning of the period, 87 percent or 1,866 remained Type I's in 2011. However, 135 of them became Type II, 105 became Type III, and 30 became Type IV (reading across horizontally from "Type I" in the left column; the data did not allow us to distinguish Type Vs). Of the 267 that were Type II, 48 ceased operating and became Type I, 159 remained Type II, 45 advanced to Type III, and 15 acquired sawmills and became Type IV. Of the 438 Type IIIs, 54 had downward mobility and ceased operating and became Type I, 135 went down a level of vertical integration to Type II, 225 remained at the same level, and 24 acquired sawmills. Finally, of the 126 Type IV communities in the sample, 6 ceased operating entirely, 12 reverted to selling logwood (Type II), 24 reverted to extraction (Type III), and 84 remained Type IV.

To summarize, there were a total of 108 Types II, III, and IV who went out of business entirely (see Type I column), most of them in forest communities in tropical areas. As we will see in chapter 7, the ecology of forest management in tropical areas has special challenges, which are reflected in these numbers. However, of all those that achieved some level of vertical integration, the total numbers of Types II, III, and IV went from 831 to 993, a gain of 162. Breaking out the numbers, the Type IIs gained, going from 267 to 441; the Type IIIs lost ground, going from 438 to 399; and the Type IVs gained, going from 126 to 153 (see figure 15). So, while there is some flux, the sector showed considerable stability with a tendency for increases in vertical integration. There is likely some relationship to the CONAFOR subsidies for this tendency, although they did not serve to arrest the many cases of downward mobility.

An analysis of which of the thirteen CONAFOR programs had the most impact on stability or upward mobility showed that support for FMPs, direct investment in equipment and machinery, internal regulations, and participatory

TABLE 4. TRANSITION MATRIX OF UPWARD AND DOWNWARD MOBILITY AND STABILITY IN TYPOLOGY FOR CONAFOR BENEFICIARIES, 2004–2011

TYPOLOGY 2004						TYPOLOGY 2011	
	Total	Type I	Type II	Type III	Type IV	Total	
Type I	2136	1866	135	105	30	1974	Type I
Type II	267	48	159	45	15	441	Type II
Type III	438	54	135	225	24	399	Type III
Type IV	126	6	12	24	84	153	Type IV
	2967					2967	

Source: CONAFOR (n.d.).

rural evaluation were all important. Statistically significant control variables included size of the community, percentage of the territory with forest, and percentage of the territory in village areas, suggesting the importance of scale. The conclusions point to the importance of minimum sizes of forest area and communities that support diversification of economic activities (Torres Rojo and Amador Callejas 2015b).

These studies show that there is a constant flux in the vertical integration hierarchy, with communities both making the decision to vertically integrate, frequently with government support, or that internal organizational issues or high transaction costs drive a decision to deintegrate. The figures also suggest that CONAFOR's efforts at launching new CFEs and pushing others up the ladder of vertical integration have had a modest degree of success and for this period resulted in net positive gains. E. Fernández Vázquez and N. Mendoza Fuente (2015) have argued that there is a sustained disappearance of CFEs and that that is one factor in a perceived "crisis" in the sector due to overregulation. There is no evidence for this assertion before 2012. However, the reduction of support during the Peña Nieto and López Obrador governments may well have led to increases in deintegration and closed businesses, but further research is needed to establish more precise numbers and the causes. Nonetheless, it is also true that the vast majority of CFEs continue to function as they have for decades, suggesting that the sector is under some stress but continues to show resilience.

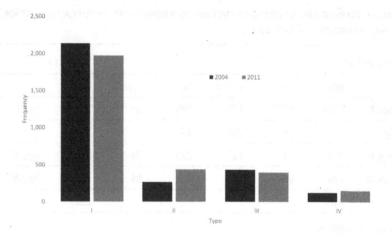

FIGURE 15. Transitions in Types I, II, III, IV (2004–11). Graph courtesy of author.

ENTREPRENEURIAL DIVERSIFICATION

In the mid-1990s the PROCYMAF program played an important role in diversifying forest-based community enterprises, focusing on ecotourism and spring-water bottling but also on pine resin harvesting, artisanry, orchid cultivation, and wildlife management, among other things (Rodríguez Salazar, Martínez, and Bray 2015). From 2006 to 2012 CONAFOR financed 223 studies for eco-tourism, water bottling, and wildlife management (Anta Fonseca 2015), though it is not clear how many of these endeavors actually produced new community enterprises. The National Commission for the Development of Indigenous Peoples (Comisión Nacional para el Desarrollo de los Pueblos Indígenas; CDI) and a predecessor government agency have also been very active in supporting ecotourism in indigenous communities in the last two decades (López Pardo and Palomino Villavicencio 2008). Ecotourism is a challenging undertaking for many communities, since, as a service industry, it requires anticipating the needs of clients who may have high expectations for a relative degree of attentiveness to their comfort. There is little literature on the extent and impacts of the efforts to promote community-based ecotourism, but they appear to have been notable. Figures from 2006 from the government tourism agency showed that 74 percent of 1,239 enterprises engaged in nature tourism were community based. These community-based ecotourism operations have emerged all over the country, but forest-based community ecotourism has been strongly documented in central Quintana Roo (Cornejo 2004; Palomino Villavicencio and López Pardo 2011) and in the Sierra Norte of Oaxaca. The Sierra Norte of Oaxaca may have one of the largest clusters of community-based ecotourism destinations globally, comprising nineteen communities that offer services (Fuente Carrasco and Ramos Morales 2013). Oaxaca has received substantial funding from the state: the Secretary of Tourism invested about US$2.7 million between 2005 and 2008, and the CDI invested US$3.6 million dollars between 2006 and 2008 in infrastructure, training, and advertising. The community of Pueblos Mancomunados (composed of eight population centers owning the same territory) was the Oaxaca pioneer in community-based tourism, beginning efforts in the mid-1990s. In 2002 I had the opportunity to nominate Pueblos Mancomunados' ecotourism operation for an award by *Conde Nast Traveler* magazine, which it won as "Best Ecotourism Destination" (Klivak 2016). Pueblos Mancomunados in 2015 received some 13,000 visitors and, due to internal conflicts about logging, is now focusing more on ecotourism than on forest management. Its spring-water bottling business has also gained the second largest market share in the city of Oaxaca, marking a transition from a logging community to one focused more on NTFPs (Bray 2016). As noted earlier, there are also now a number of forest communities who have CFEs based only on NTFPs. For example, a study of twenty-five communities with forests in the Mixteca region of Oaxaca found that six produced only NTFPs, four for pine resin and two for bottled water. Several communities in the sample were also exploring ecotourism (Hernández-Aguilar 2017). These are communities that merit a Type VI designation as CFEs, as discussed earlier.

FROM INDUSTRIAL VERTICAL INTEGRATION TO INTERCOMMUNITY COLLECTIVE ACTION: THE ROLE OF FOREST ASSOCIATIONS

In addition to vertical integration decisions, Mexican CFEs also commonly make decisions as to whether or not to engage in intercommunity collective action. As J. Fox (1996) has noted for Mexico, dense social capital at the community level can be highly segmented spatially. Neighboring communities frequently have histories of conflict and tension over territories and resources, and more distant communities can also inspire great mistrust. The emergence of intercommunity collective action around economic issues is not the norm, since it must "overcome the socially constructed constraints of locally confined solidarities" (Fox 1996, 1091). However, despite a few examples (García-López and Antinori 2018; Ojha et al. 2016; Paudel, Monterroso, and Cronkleton 2010), intercommunity collective action has received little consideration in the literature. As noted in chapter 1, Ostrom came to appreciate the importance of intercommunity organizations in her comment that "modest coping method for dealing with threats to sustainability" include "the creation of associations of community-governed entities" (2005, 279), in this case forest associations. These associations are frequently turbulent in their functioning and require both the prior existence and strengthening of trust and social capital (Bray et al. 2013). As in other aspects of community forestry in Mexico, the national government has played a key role in providing incentives for intercommunity associations, such as the legislation in the Echeverría period that required the formation of ejido unions. Klooster (1997, 43) has emphasized the role of the Mexican state in promoting intercommunity collective action, noting that "in contrast to the influential description of millennial accretion of social capital described for Italy. . . . activist elements in central government can jump-start the process by creating participatory associations and user-groups which circumvent exclusionary elements of local governments and rural elites."

It has been a notable characteristic of community forestry in Mexico that the majority of CFEs have, at one time or another, participated in FAs at both the regional and national levels. The existence of at least forty-three FAs in Mexico have been recorded in the literature, though the fate of most of these organizations has not been well documented (Bray and Merino-Pérez 2003). In a survey of forty-one communities from Michoacán and Durango, 78 percent currently belonged to an FA at the time of the survey; participation rose in step with vertical integration (Antinori and Rausser 2010).

Most forest associations have been encouraged or created by the state, but some originated with combinations of grassroots mobilization and top-down state initiatives, the strength of each influence varying historically and within the same organization. "Rather than purely self-organized or entirely state-led[,] . . . multi-level forest commons . . . have emerged out of hybrid and evolving combinations, structural conditions, and creative collective agency, merging conflict and cooperation" (García-López and Antinori 2018, 201). Whatever their origins, forest associations are important elements in multi-level or networked governance (García-López and Antinori 2018) and can bring economic efficiencies such as shared forest technical services and channeling of government support. Forest associations in Mexico provide political representation, environmental programs, information on prices, monitoring of logging contracts, marketing services, silvicultural plans, and training and channeling of resources—all important economic functions. The intercommunity organizations in Quintana Roo formed by the Plan Piloto, and the first generation of associations in Durango were able to organize what were essentially price cartels, setting prices for their timber (which, as noted earlier, Ostrom suggested was not a characteristic of common property regimes). Forest associations have also engaged in collective action around the establishment of tree nurseries, marketing strategies, equipment, and infrastructure for firefighting, roads, and communications (Torres Rojo and Graf Montero 2015). The Union of Forest Ejidos of the Sierra Norte of Puebla (Unión de Ejidos Forestales de la Sierra Norte de Puebla), whose average forest size is just over 200 hectares, jointly operated a sawmill for decades, though it stopped operating more recently. There are positive correlations between association membership and investments in secondary processing, forest diversification, local public goods investments, control of illegal logging, self-reported improvements in forest cover, and presence of wildlife (García-López and Antinori 2018).

Intercommunity collective action from the grassroots has also occurred at the level of an entire state. The Coordination of Forest Organizations and Ejidos of Oaxaca (Coordinadora de Organizaciones y Ejidos Forestales de Oaxaca; COCOEFO) was an unusual case of a state third-level organization. It was founded in the early 1990s, supported by the NGO ASETECO, and had some success in a struggle against the federal Treasury Department to exempt communities from enterprise taxes, arguing that the enterprises supported many things that were normally the responsibility of government (ASETECO 2002). In the early 2000s, the first major government effort since the 1980s to launch new forest associations occurred. The 2003 Forest Law required CONAFOR to create Regional Forest Management Units (UMAFORs) at the watershed level in each state, including the grouping of regional Silvicultural Associations into state-level associations (García-López 2013). CONAFOR created PROFAS in 2004 to provide subsidies to these new regional and state forest associations, as discussed in chapter 3. However, PROFAS only supported the new organizations, which were to also include small private forest owners, and did not support the many existing ejido unions and other intercommunity organizations, raising criticisms that it was an effort to reinstitute state corporatist control of the forest sector. However, by 2011 government policy had changed to deemphasize the UMAFORs and supported all existing associations in a region. In general, the development of the regional and state Silvicultural Associations has been erratic and completely dependent on government funds (Torres Rojo and Graf Montero 2015).

In addition to associations based in regions or in states, there have been efforts over the decades to promote national forest associations. As with regional ones, the national organizations have been supported or created by government agencies, even those that initially emerged as grassroots mobilizations. Three different national organizations have emerged in different periods: (1) one in the late 1980s, promoted by the forestry subsecretary with some later involvement by the CNC, (2) the Mexican Network of Peasant Forest Organizations (Red Mexicana de Organizaciones Campesinas Forestales; RED MOCAF) formed in 1992 as an offshoot of a national campesino organization that has been the most enduring of the national organizations, though it does not focus exclusively on forestry, and (3) The Union of Community Forest Organizations (Unión de Organizaciones Forestales Comunitarias; UNOFOC) created in 1995 supported by another government agency, and currently with a low level of activity (Bray and Merino-Pérez 2004). These national organizations, which had few economic functions and served mostly as political lobbying groups and to some degree channeled government support, faltered when support declined or ended. However, the following section will examine two notable examples of entrepreneurial forest associations or alliances among CFEs for production.

COLLECTIVE ACTION AND INTERCOMMUNITY ENTREPRENEURIAL ALLIANCES

The first and also very impressive example of intercommunity entrepreneurial alliances is that of the Union of Forestry and Agrarian Ejidos and Communities "General Emiliano Zapata" (UNECOFAEZ), established in northwestern Durango in 1976 and encompassing twenty CFEs.[7] The history of UNECOFAEZ until the late 1980s is discussed in chapter 2, and even then it was the community forestry organization with the largest area of influence in Mexico (G. Chapela y Mendoza 1998). As of 2019, it included forty-six ejidos and

---

7. In addition to the citations, information in this section is based on interviews with Grupo SEZARIC administrators, Santiago Papasquiaro, February 2019.

twenty-nine comunidades, covering nearly 1.1 million hectares of territory, its forest technical services provided by the UAF Santiago Papasquiaro. It operates a water-bottling plant, a restaurant, a warehouse for agricultural supplies, a radio communications systems, a machinery workshop, and other industries. However, UNECOFAEZ's most noteworthy entrepreneurial alliance is in the operation of a plywood factory and associated industries, a factory obtained through applying political pressure to PROFORMEX in 1989. By 1990, with support from the federal government and against the opposition of the state governor and state timber industrialists, the union was able to lease and later purchase PROFORMEX's plywood factory, forming the General Emiliano Zapata Silvi-Industrial Group (Grupo Silvindustrial General Emiliano Zapata; SEZARIC) to administer it. However, the factory was in such deteriorated shape, it was not operational. In 1991, as a result of a presidential visit, the group was able to acquire a subsidized federal loan to rehabilitate the factory, and SEZARIC paid back the loan by 1993. Governance of such a large organizational structure involving so many communities has commonly been turbulent, including frequent leadership struggles and tensions in the social enterprise between social and economic goals. In response to both internal conflicts and external challenges, the union pursued a high-risk strategy of diversification and expansion in the 1990s, using both its own capital and subsidies from federal government agencies. In the mid-1990s it established a credit union, a marketing firm to buy timber for the plywood factory, a furniture factory, and a center to train leaders and sawmill administrators. It also diversified into nonforest activities, with its own extension agents for agriculture and livestock raising and other development projects (Taylor 2000). By 1999 the plywood factory had forty community shareholders (not all of the members of UNECOFAEZ are associates), and the Union also administered a machine tool shop, five forest technical service offices, a plant nursery, the credit union, an agricultural inputs store, seven road improvement committees, and the training center, most of these activities receiving state subsidies (Taylor 2001). All of this entrepreneurial activity generated frequent turbulence; the union's interest as a buyer conflicted with the interests of the member organization as producers, and the union received accusations that it had become a "timber buyer like any other" (Taylor 2000, 263). Its complex set of activities meant that the union became a sort of holding company, where "each entity keeps its own books and has its own administrative council." As one participant described it, "The Union is like a game board with pieces. They seem independent, but they're going in a single direction. In a crisis, they come together" (Taylor 2000, 268; Taylor 2001).

Problems persisted surrounding participation of the member communities, and many communities felt distant from the operation of "their" organization. These tensions have persisted over the decades but continue to be successfully managed as of 2019. The union also has the ongoing challenge of many of the member communities' having an uneasy coexistence with organized crime. In 2018, Grupo SEZARIC had total sales of US$16.6 million with around 8 percent profitability, and it distributed some 60 percent of the profits to community shareholders. The union has 550 direct employees, 40 percent of whom are women, who are paid the same as men, and some of whom have physically demanding jobs, and they are also hiring individuals with disabilities. Some 20 percent of the union's production is FSC certified, and it plans to increase that in coming years. It has also recently constructed a biomass cogeneration plant that represents an investment of some US$3 million, considered a pioneering project in Mexico ("Primera planta" 2020). UNECOFAEZ and Grupo SEZARIC represent an extraordinary but still poorly documented example of intercommunity entrepreneurial vigor.

The second example of entrepreneurial intercommunity collective action is that of Textitlán-Ixtlán-Pueblos Mancomunados Furniture (Textitlán-Ixtlán-Pueblos Mancomunados Muebles; TIP Muebles). This furniture manufacturing and marketing alliance, which emerged in a dialectic between government support and community agency, consisted of the government opening entrepreneurial spaces that were quickly occupied by vigorous community collective action. In 2004–5 the federal and Oaxaca state governments began meeting with three of the largest CFEs in Oaxaca (Santiago Textitlán, Ixtlán de Juárez, and Pueblos Mancomunados) around a new industrial strategy to develop the forest sector in Oaxaca, expanding or helping to create furniture factories in the three communities. All three communities already had FSC certification for well-managed forests and obtained chain-of-custody certification to produce certified furniture. The state, with some support from the U.S. Agency for International Development, provided new physical, financial, and human capital, and the new factories also got a very significant boost from the Oaxaca State government through a solidarity/green purchasing policy on the part of the Oaxaca Institute of Public Education (Instituto Estatal de Educación Pública de Oaxaca; IEEPO). The procurement policy that all new desks for public schools in the state be made of certified timber led to orders totaling some 116 million pesos between 2006 and 2009. However, a 2006 prolonged strike by a teacher's union delayed payments and diminished orders from IEEPO, spurring a search for new marketing strategies not as dependent on government purchases (Klooster, Taravella, and Hodgdon 2015; Tanaka 2012).

Thus, in 2006 TIP Muebles was formed. The Communal Forest Integrator of Oaxaca (Integradora Comunal Forestal de Oaxaca, S.A. de C.V.; ICOFOSA), which was the entrepreneurial alliance that established TIP Muebles, coordinated joint production and marketing of furniture and launched three retail stores in Oaxaca City (Klooster and Mercado-Celis 2016). The concept of "Integradores" also responded to a larger government strategy at the time to encourage the many individual carpenter shops in several CFE communities to band together to try and achieve efficiencies of scale and larger markets (Klooster, Taravella, and Hodgdon; Tanaka 2012). TIP Muebles has suffered from many of the problems of the member CFEs: limited marketing and design capabilities; high production costs, due to lack of training; high turnover of administrators, due to traditional governance structures; and lack of management skills. However, a substantial degree of new social capital was also created by the construction of the entrepreneurial alliance among the three communities; evidence for this is the response after a fire destroyed Pueblos Mancomunados' sawmill and furniture factor: Ixtlán de Juárez quickly volunteered its spare manufacturing capacity to continue operations. The increased trust and social capital arising from this support are what led to TIP Muebles opening new franchise stores in Mexico City and Puebla within eighteen months after the fire (Villavicencio Valdez, Hansen, and Bliss 2012).

ICOFOSA allowed the three CFEs to learn the furniture business, diversify markets to the general public and government institutions, and organize joint responses to large orders. The stores are not required to be economically self-sufficient, and monthly subsidies from the partners make up deficits, though the furniture production itself has been found to be modestly profitable (Tanaka 2012). A survey of purchases by furniture wholesalers from TIP Muebles found that 95 percent would buy from TIP Muebles again. As well, the alliance was valued as a symbol of the "Oaxacan community-based green economy" by the government procurement agencies (Tanaka 2012, 29). However, as of late 2018, after over a decade of relative success, the model had run into significant problems. Santiago Textitlán, after closing its furniture factory in 2017 due to lack of profitability, ceased participating in the alliance. Moreover, an ambitious expansion plan that led to eight stores throughout Mexico failed, leading to a retrenchment back to two stores in Oaxaca City supplied only by two of the original communities, a serious challenge to the business model. TIP Muebles is currently reevaluating its strategies (interview with I. Santiago, December 12, 2018); it is not clear whether this model has a future.

Mexican CFEs are unique market-oriented and social enterprises and require an extension of the theory of the firm to account for them. Despite multiple deficiencies as businesses and intense international competition, the safe political, social, and economic operating space that was opened by top-down reforms and vigorous community response has allowed them to use their natural capital, the other four capitals, and a high-quality product to successfully compete in national markets. The relatively high, if fluctuating, prices for timber have allowed communities to pursue objectives other than profit maximization. Generation of employment, public goods, and intergenerational values emerge as alternatives to profit maximization. The uniqueness of Mexican CFEs as firms and social enterprises is based on a community owning a common property resource fully supported by the Mexican legal framework. State policy and program support has enabled the community forest sector to remain relatively stable and even reduce poverty levels. These enterprises make choices about vertical integration, engage in intercommunity collective action through forest associations, and enter into entrepreneurial alliances that seek to maximize their multiple objectives. These community timber businesses have been able to survive in the marketplace and remain profitable, even highly profitable, due to their access to a high-quality timber resource that the marketplace values. These successes reflect the resilience of the CFE sector.

CHAPTER 7

PUTTING THE ECOLOGY IN THE SOCIAL-ECOLOGICAL SYSTEM

This chapter turns to the ecological dimensions of the Mexican community forestry SES. Mexican forests are ecologically important for multiple reasons: in addition to timber production, they harbor much of Mexican biodiversity and provide ecosystem services in hydrology, soils, and carbon capture, among other functions (Perez-Verdin et al. 2018), and they play an increasingly important role in ecotourism. Researchers suggest that up to 12 million Mexicans live in and depend on forests for parts of their livelihoods (Eguiluz-Piedra 2004), though some estimates are higher. The vast majority of Mexican forests are not pristine and have been inhabited and used by human beings with varying degrees of intensity for millennia. For example, of an estimated original 93,560 square kilometers of pine-oak habitat in the Western Sierra Madre of Chihuahua and Durango, less than 1 percent (571 square kilometers) remained as old growth in 1995 (Lammertink et al. 1996). Technically, the concept of ecology refers to the relations of organisms to one another and to their physical surroundings (Begon, Harper, and Townsend 1986). For purposes of this chapter, we use "ecology" to refer to a broad vision of ecological processes in managed forests, including general characterizations of extent, management practices, and impacts on conservation status, structure, and species composition of Mexican forests. The conceptual framework presented in chapter 1 proposed that the ecology of Mexican forests has been shaped since the Mexican Revolution by state policy, markets, and community collective action and that these social dimensions have existed in intimate coupling with the forests. Specifically, as the framework indicates, the practice, both informal and formal, of land use zoning has provided the structure that strongly influences, even determines, forest cover processes (including deforestation, forest maintenance, and forest recovery), silvicultural practices and certification, environmental services, and biodiversity conservation.

Against this background of ecological and land use patterns, this chapter reviews forest ecology and biogeography and the formal land-use zoning programs represented by the FMPs and Community Territory Land-Use Zoning exercises (Ordenamiento Territorial Comunitario; OTCs) that limit anthropogenic disturbances to particular areas. The chapter will then examine how these zoning practices and CFE management in general have influenced (1) deforestation, the most devastating consequence for forest ecology, and the possible role of CFEs in braking it, as well as associated processes of forest maintenance and forest recovery; (2) silviculture practices as a potentially relatively benign anthropogenic disturbance of the ecological system (Oliver and Larson 1996) and the role of third-party certification of sustainable forest management; and (3) state conservation initiatives such as the Voluntary Conservation Areas (Áreas Destinadas Voluntariamente a la Conservación; ADCVs) and Payment for Hydrological Services programs (PHS), both of which provide varying subsidies for environmental services and biodiversity conservation. These conservation zoning practices commonly give only an extra layer of protection over preexisting community land-use patterns. We argue that these practices have had implications for forest productivity and forest conservation that have driven positive feedback to state policy, markets, and community governance and that hence amplify policies and practices in support of CFEs, as specified in the conceptual model presented in chapter 1.

The geographic distribution and most important forest types in Mexico were reviewed in chapter 2, but we will briefly recap and expand on that discussion. Map 1 in chapter 2 shows the historic or potential forest cover in Mexico, while map 2 here shows the current distribution of the major forest types in Mexico.

Temperate and tropical forests account for 65.3 million hectares or 33.57 of the Mexican national territory. Of this, 51.7 percent is temperate forests (33.8 million hectares), and 48.3 percent (31.5 million hectares) includes humid, semihumid, and dry tropical forests. Of this, it is estimated that 60.1 percent of the temperate forests (20.3 million hectares) are in 9,248 communities and

Current vegetation

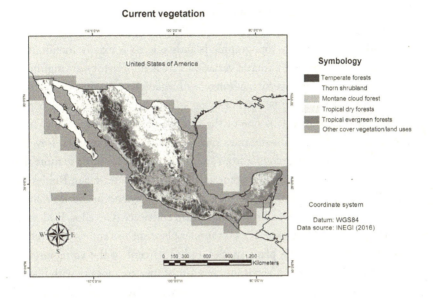

MAP 2. Current vegetation in Mexico. Coordinate System: Datum—WGS84. Data source—INEGI (2016).

61.6 percent (19.4 million hectares) of the tropical forests are in 11,847 communities, for a total of 21,095 communities with some amount of forest (Torres Rojo and Amador Callejas 2015a). As noted in chapter 3, estimates of the percentages and total hectares of forests in community territories vary from 58.8 percent to 61.8 percent, while the total forest area varies from 38.5 million to 48.6 million (Madrid et al. 2009; Skutsch et al. 2013, Torres Rojo and Amador Callejas 2015a), so the numbers used here should be taken under advisement. Significant portions of forests on community lands are not commercial. Of the 20.3 million hectares of temperate forest, some 46 percent are mostly noncommercial oak forests, and of the 19.4 million hectares of tropical forests, 52 percent are noncommercial dry tropical forest. The total communities include those who have mixed forest types in their territory, such as various combinations of temperate and tropical forest and other forms of noncommercial vegetation. Communities with pure stands of one forest type tend to be heavily forested, with an average of 66.3 percent forest cover in temperate forest communities and 60.7 percent in tropical forest communities (Torres Rojo and Amador Callejas 2015a, calculated from table 1.5, 26).

These forests also contain much of Mexico's biodiversity (Sarukhán and Dirzo 2001; Sarukhán et al. 2014). The meeting of the Nearctic and neotropical biotic regions, many topographic islands due to Mexico's mountainous landscape, and the wide climatic variation across its territory are significant factors in Mexico's biodiversity. Mexico is a megadiverse country estimated to contain approximately 10 percent of living species, though it is only 1 percent of the global land area. It is one of the top five countries in species richness of vascular plants and of vertebrates such as mammals and reptiles. Levels of endemism are also high, ranging from around 10 percent for birds to more than 60 percent for amphibians and some plant groups (Mittermeier, Robles Gil, and Mittermeier 1997). Its mammalian fauna ranks second in species richness globally, and 32 percent of all its mammals are endemics (Ceballos, Rodríguez, and Medellín 1998). Of some 25,000 vascular plant species and 1,352 vertebrate species, 81 percent of the plant species and 75 percent of the vertebrates are found in the forests. Most Mexican forest landscapes are mosaics with historic legacies of use that have contributed to diversity. Biodiversity is particularly rich in tropical and mountain cloud forests (Toledo 2010), and temperate forests are also rich in tree species compared to temperate forests further north (Farjon and Styles 1997). Public protected areas are the preferred institutions for many to protect this biodiversity, but one study found that only 54 percent of Mexican protected areas are effective, whereas the other 46 percent are weakly effective or noneffective (Figueroa and Sánchez-Cordero 2008). On the other hand, a recent study found protected areas in the Yucatán Peninsula to be generally effective, reducing deforestation by 8–12 percent (Miteva et al. 2019). In any event, while there are 20.7 million hectares in terrestrial protected areas, there are some 39 million hectares of forest on community lands, making them key for forest biodiversity conservation in Mexico (SEMARNAT/CONANP 2016; Torres Rojo and Amador Callejas, 2015a). In addition, realistically there are not enough resources to expand, manage, and maintain more protected areas, and community conservation will be much more cost-effective.

The two most important forest ecosystems for the temperate forests for logging are the pine and pine-oak forests (or coniferous forests, *bosques de coníferas*) and for the tropical forests, humid and semihumid forests (*selva alta perennifolia* and *selva mediana subcaducifolia*, with other subcategories). Communities are estimated to have 10.98 million hectares of coniferous forests for commercial purposes and 8.73 million hectares of tropical forest that are commercially valuable. Based on these figures, a total of 19.71 million hectares of forests with commercial potential are on community lands. J. M. Torres Rojo and J. Amador (2015a) suggest that there are 6.2 million hectares under FMPs in community lands. This information suggests that there may be a total of 19.71 million hectares of forests on community lands, around 31 percent of which is under management for timber, figures that have important implications for forest conservation.

LAND USE ZONING

THE FOREST MANAGEMENT PROGRAMS (FMPS)

The social processes of the SES created a protected economic sphere in which CFEs could develop. The SES also resulted in a standardized top-down set of management rules that constrained forest cover processes and silvicultural practices and provided for environmental services and biodiversity conservation through land use zoning across all community forests. The contribution of the community forestry SES to forest cover, forest management, and biodiversity conservation in Mexico occurs through multiple pathways. The principal formal institutions for land use zoning are the FMPs that as of the 2003 Forest Law, zone the entire community territory into four broad categories: (1) conservation areas, (2) forest production areas, (3) forest restoration areas, and (4) other uses, mostly agriculture and village areas (Bray and Duran-Medina 2014). Communities frequently only manage small percentages of their forests, dedicating substantial forest entirely to conservation, though the amount designated varies with the total size of forest. J. Barsimantov and J. Kendall (2012) study of 733 municipalities in eight states suggests high percentages of forests in CFE communities are not logged; that is, 62.8 percent of coniferous forests in these municipalities are owned by communities, but only 9.6 percent are being logged. The F. W. Cubbage et al. (2015) study of thirty CFEs in twelve states with a mean of 12,269 hectares found much lower but still substantial areas of forest in conservation, that is, 72 percent of the forests in production and 28 percent in conservation. A study of twenty-three CFE communities in the Sierra Norte of Oaxaca found that in a total area of 201,093.94 hectares, 37 percent of the total (74,293 hectares) were under various forms of conservation and 78 percent were forested. The forest areas included dry tropical forests, pine-oak forests, and cloud forests, and a clear relationship existed between larger forests and larger conservation areas (Pazos-Almada and Bray 2018). A study of the FMPs for fifty-nine CFEs in Chihuahua found that 28.2 percent of their forests were under conservation (Bray and Duran-Medina 2014). A noteworthy example of conservation is Ixtlán de Juárez in Oaxaca, which protects nearly 6,000 hectares of uninhabited tropical montane cloud forest, which would be a wilderness area by U.S. definitions (Pacheco-Aquino, Durán Medina, and Ordóñez-Díaz 2015; Van Vleet, Bray, and Durán 2016).

The FMP land-use zoning is based on the regulations of the Forest Law of 2003 with greater detail in the NOM-152 (SEMARNAT/CONANP 2006) approved in 2008. This land use zoning also appears in a number of cases codified in community written statutes. The FMPs set up land use zoning into the four broad categories mentioned above and use a standardized methodology, required in the NOM-152, that includes using remote sensing and GIS technologies. The zoning does not necessarily create uniform large blocks of land use. Agriculture, for example, can be scattered around in smaller plots that have favorable soils and other conditions. The forest masses are generally in large but not always contiguous blocks. The conservation areas classified in the FMPs can include a variety of categories, such as flora and fauna habitat protection, riparian buffer zones, slopes greater than 100 percent or 45 degrees, areas over 3,000 meters above sea level, cloud forests, recreation areas, and areas of high conservation value. Some of these conservation areas, such as riparian buffer zones, are within the production forest areas (Pazos-Almada and Bray 2018). The land use categories are generally not viewed as an imposition, but merely ratify long-standing patterns and norms of land use with deep roots (S. Anta, personal communication, April 19, 2018).

For example, in the Quintana Roo community of X-Maben, conscious conservation rules stem from at least the 1950s, when the community prohibited forest conversion in areas where the *Manilkara zapota* tree was dominant, valuable for its chicle. The PFAs for timber production established with the PPF in the 1980s were largely placed over the existing informal community chicle reserves, a new commercial incentive for forest conservation (Dalle, Pulido, and Blois 2011). Another example of deeper roots of forest conservation lies in the comparison between communities in and around the Calakmul Biosphere Reserve in the state of Campeche, Mexico, and the Maya Biosphere Reserve in the Petén of Guatemala. Bray et al. (2008) proposed that communities with CFEs had lower deforestation in both regions; however, C. R. Solorzano and F. Fleischman (2018) argue that the generally much higher rates of conservation around the Calakmul Biosphere Reserve are due to earlier institutions that pre-date reserve establishment and the establishment of CFEs. The agrarian reform system in Mexico, the researchers contend, resulted in lower population density, a higher degree of tenure security, and more economic and political equality in the Calakmul Biosphere Reserve, factors that influenced communities there to have more positive attitudes toward conservation and to create more conservation reserves.

COMMUNITY TERRITORIAL ZONING (OTCS)

The OTC methodology was developed in Sierra Norte of Oaxaca by the NGO Estudios Rurales y Asesoría (ERA) in the early 1990s as a reaction to top-down ecological planning proposals from the Mexican government (ERA/UZACHI 1994). OTCs are commonly finer grained and more participatory than the FMPs (Anta Fonseca et al. 2006). The method, first implemented in sixteen communities and watersheds in Oaxaca covering 220,003 hectares, placed under regulated community conservation almost a quarter of that territory (F. Chapela y Mendoza and Y. Lara 2007). F. Chapela y Mendoza (2006, 57) suggests significant impacts of the exercises: "It is frequent that the agrarian communities that become involved in these planning processes change their perception of their options for development, not because a technician or outside official 'raises their consciousness' about the importance of conservation, but because their members undertake the collective conformation of a vision of the whole of their territory and imagine a future based on taking advantage of their resources."

The methodology was adopted by PROCYMAF in the 1990s and became one of their important program instruments. From 2006 to 2012 a successor program to PROCYMAF in CONAFOR supported 924 OTCs covering 6.2 million hectares (Anta Fonseca 2015). This program included both communities who had FMPs and a significant number that did not, thereby notably expanding conservation and sustainable use planning of Mexican community forests beyond timber production. There have been no evaluations of the actual impact of this large number of OTCs but F. Chapela y Mendoza (2006, 57–58) argues that the model creates a culture of community territorial planning that "establishes a framework of long-range planning and a structure of agreements, rules, sanctions, and levels of authority that make it possible to transparently harmonize divergent . . . interests." The OTCs are also commonly used to establish written community rules governing territorial use and other aspect of community life. Oaxaca has been the most important state in community conservation, both formal and informal. G. J. Martin et al. (2011) found that there were 126 formal and informal conservation areas in Oaxaca covering 375,457 hectares, 14.5 percent more than the public protected areas in the state.

DEFORESTATION, FOREST MAINTENANCE, AND FOREST RECOVERY

The rise of community forestry in the 1970s coincided with the most intensive period of deforestation in Mexico, the heaviest deforestation occurring in the tropical areas. Deforestation rates in tropical forests in southeastern Mexico, where they are concentrated, occurred at alarming rates beginning as early as the 1950s. The northern Lacandón forest had an annual rate of loss of 12.4 percent in the 1960s (Bray and Klepeis 2005), possibly the highest ever recorded in Mexico. In the mid-1980s deforestation rates in all tropical forests were estimated at around 2 percent a year (World Bank 1995), and a study of the eight states of southeastern Mexico concurred, reporting a 1.9 percent loss per year from 1977 to 1992 (Cairns et al. 2000). For the state of Oaxaca, higher deforestation rates in tropical areas (around 2 percent annually) than in temperate forests (around 1 percent annually) were confirmed for the 1980–2001 period (Velázquez et al. 2003). A study of dry tropical forests found only 27 percent of the original cover intact by 1990, estimating an annual deforestation rate of 1.4 percent with heavy fragmentation and disturbances (Trejo and Dirzo 2000). However, for all of southern Mexico, Vaca et al. (2012) found that the annual rate of deforestation fell from 0.33 percent in 1990–2000 to 0.28 percent from 2000 to 2006; forest loss stayed concentrated in a few hotspots, and a very modest forest recovery of some 0.20 percent per year took place, mostly as a result of agricultural abandonment.

In recent decades forest loss has always been higher in tropical forests than in temperate ones. The 2000 National Forest Inventory suggested that the average overall rate of deforestation for 1976–2000 was 0.43 percent, but the rate of loss in tropical forest was 0.75 percent, while temperate forest loss was only 0.25 percent. This implies an average annual loss of 86,718 hectares for temperate forests and 263,570 hectares for tropical forests, for a total average annual loss of 350,288 hectares (Velázquez et al. 2000). According to official figures, deforestation in

Mexico had been declining from previous high points until the early 2010s. From 1990–2000, the rate of loss was estimated at 354,000 hectares annually (0.52 percent), declining to 235,000 (0.35 percent) from 2000 to 2005, to 155,000 hectares (0.24 percent) from 2005 to 2010, and to 92,000 hectares annually from 2010 to 2015 ("Consulta temática" 2015), though these numbers vary in different sources. This last rate of deforestation is only 26 percent of the rate in the 1990s. The reduction in deforestation was particularly dramatic in the commercial temperate forests. In pine forests, forest loss was an average of 29.498 hectares from 1993 to 2002, but net zero deforestation had been achieved by 2002–7, with an average annual forest gain of 928 hectares. In pine-oak forests, deforestation declined from an average of 35,190 hectares from 1993 to 2002 to 8,889 hectares from 2002 to 2007, with forest loss in oak forests even less (CONAFOR 2010). By 2002–7 deforestation was almost entirely confined to humid, semihumid, and dry tropical forests, though even here rates had declined notably from the earlier period. Then, in a troubling turn of events, deforestation reportedly jumped dramatically under the Peña Nieto government (2012–18) from a baseline of 158,000 hectares to 299,000 hectares annually, principally in the tropical forests of Chiapas and Campeche (Benet 2018). Setting these most recent numbers aside, what, if anything, did CFEs have to do with the historical decline until the most recent period?

Deforestation in often presented as a linear process, and, indeed, in tropical areas of Mexico a major pulse of deforestation occurred from the 1950s to the 1980s. However, forest cover processes have usually been more complicated and nonlinear. Two major pathways for forest cover processes in Mexico and Central America have been proposed: (1) the continued deforestation pathway and (2) the forest recovery, maintenance, and protection pathway, the latter including public protected areas, community conservation, sustainable forest management, and shade tree coffee (Bray 2010b). From the 1950s to the 1980s, deforestation indeed happened usually in a linear process. However, beginning in the 1990s some regions of Mexico showed continuing deforestation, whereas others showed forest recovery, and still others—such as areas of shifting agriculture, coffee agroforestry, some public protected areas, and community forests with CFEs—were stable or even expanding in overall forest cover, albeit within a shifting mosaic of forests, forest recovery, and production areas. Although tropical deforestation remains higher than temperate forest deforestation, there are also tropical regions where deforestation is declining and temperate regions where it is climbing (Bray 2010b).

For example, the temperate forests in the region of the Monarch Butterfly Biosphere Reserve in highland Michoacán and Mexico State showed rapid deforestation over the last several decades due to heavy illegal logging and subsistence agricultural expansion. From 1971 to 1984 the deforestation rate for the reserve region was 1.7 percent annually but accelerated to 2.41 percent from 1984 to 1999 (Brower et al. 2002). Deforestation there, however, had been substantially controlled by the 2010s (Vidal, López-García, and Rendón-Salinas 2014; Vidal and Rendón-Salinas 2014). The temperate forests of highland Chiapas, where there are no CFEs, also showed intensifying deforestation in the 1990s (Cayuela Benayas, and Echeverría 2006; Ochoa-Gaona and González-Espinosa 2000). In contrast, some of the tropical forest hotspots have shown drops in deforestation rates in recent periods. In the Lacandón rainforest, historically the demographic escape valve for highland Chiapas and other areas of Mexico, the rates of forest loss declined from 2.13 percent annually from 1974 to 1984 to 1.6 percent for 1984 to 1991 (Mendoza and Dirzo 1999). Deforestation also declined by 1997 in the region of the Calakmul Biosphere Reserve in Campeche (Roy Chowdhury and Schneider 2004). The decline in deforestation rates in tropical areas is due essentially to the end of large-scale directed tropical colonization and the discouragement of spontaneous tropical colonization through the declaration of protected areas. For example, southern Campeche and Quintana Roo were the site of large-scale directed colonization projects in the 1970s and 1980s, but no new ejidos were established in the region of Calakmul after 1991. Vaca et al. (2012) also found that for all of southern Mexico, forest cover change was complex and nonlinear in the 1990–2006 period and suggested an incipient forest transition as of the mid-2000s. On the other hand, some tropical regions of Mexico, such as the important Grijalva-Usumacinta watershed, continue to show high rates of forest degradation and deforestation of secondary forests (Kolb and Galicia 2012).

The end of colonization, ongoing outmigration, associated agricultural abandonment and regrowth, and conservation policies have all surely contributed to the decline of deforestation in Mexico until the most recent period. However, there is also evidence that community forest management through CFEs has been a significant contributing factor, and that CFE management compares favorably with public protected areas in controlling deforestation. In the tropical areas, a series of studies in the Yucatán Peninsula have shown that CFE management reduces deforestation. A study of central and southern Quintana Roo reported deforestation rates of 0.4 percent for the 1976–84 period, before the advent of community forestry in the region, and this already low rate declined to a vanishingly small 0.1 percent, for the 1984–2000 period, when CFEs became established. Further, there are no public protected areas in the region, yet its rate of forest loss is lower than any region of southeastern Mexico with a protected area (Bray et al. 2004). A comparison of central Quintana Roo with an area of northern Campeche that includes the buffer zone of the Calakmul Biosphere Reserve and is part of the Mesoamerican Biological Corridor demonstrated that deforestation was 0.7 percent from 2000 to 2005 around the Calakmul Biosphere Reserve, whereas in Quintana Roo, dominated by CFEs, there was actually a small degree of forest recovery. As well, central Quintana Roo had twice the population density, suggesting that "community forest management can play an effective role in forest conservation" (E. A. Ellis and Porter-Boland 2008, 1971).

Another recent study of deforestation in the entire state of Quintana Roo found that it was associated with agriculture, livestock production, and tourism, but in regions where there was traditional milpa agriculture and CFEs, deforestation was much lower. Better community organization and participation, the presence of investments, and higher CFE vertical integration heightened the conditions for forest maintenance and biodiversity conservation (E. A. Ellis, Romero Montero, and Hernández Gómez 2017). A study of eight ejidos in Quintana Roo showed that communities that maintained forest commons had lower deforestation rates, particularly those that had logging permits (DiGiano et al. 2013). Another study that compared community forest management regions in Quintana Roo and Guerrero with a national sample of sixty-seven public protected areas found that both the public protected areas and the community forest regions maintained over 95 percent of their original forests and there was no significant difference between them in deforestation rates (Durán, Mas, and A. Velasquez 2005).

Cross-national comparisons of the adjacent tropical forests around the Calakmul Biosphere Reserve in Mexico and the Maya Biosphere Reserve in the Petén of Guatemala also suggest that community industrial forestry compares well with protected areas for conservation while providing for sustainable local livelihoods. Bray et al. (2008) compared deforestation in nineteen community forests in Mexico and community concessions in the Guatemalan Petén and eleven protected areas in both countries that had widely varying conditions, particularly with respect to colonization pressures. Bray and colleagues found that that the deforestation rate for community forests was approximately half that for protected areas across the region, and that long-inhabited communities with forest management can be as effective as uninhabited parks at delivering forest protection and more effective at delivering local benefits. In the Petén, community concession beneficiaries have incomes that are as much as three times higher than regional averages based only on two- or three-month periods of work. It has also been suggested, as noted earlier, that differences in the degree of conservation between the two regions is due to institutional legacies of the agrarian reform that resulted in a lower population density, greater tenure security, and greater economic and political equality in the communities in and around Calakmul Biosphere Reserve, influencing these communities to have more favorable attitudes toward conservation and to create more conservation reserves (Solorzano and Fleischman 2018).

For temperate forests, a study of 733 municipalities in eight states with at least 50 hectares of coniferous forests found that municipalities with higher percentages of commonly owned forest and higher percentages of common forest under FMPs "both reduce the gross and net rates of deforestation and increase the rate of forest recovery of coniferous forests." This effect did not hold for nonconiferous forest, suggesting that forests are conserved when they are valued for their timber (J. Barsimantov and J. Kendall 2012). In Michoacán, avocados have been to the temperate forests there what soybeans have been to the tropical forests of Brazil. J. Barsimantov and J. Navia Antezana (2012) studied avocado-producing regions of Michoacán and found that ten municipalities that produce avocados lost 39.5 percent of their forests between 1990 and 2006 for a 3.1 percent annual deforestation rate, much higher than the 0.25 percent rate found for temperate forests nationally. By contrast, two case studies of communities in the region that had forest management programs showed deforestation rates of 1.3 percent and 0.5 percent, higher than national averages but substantially lower than the average for the region, suggesting some inhibiting effect from forest management. Above we saw that the region of the Monarch Butterfly Biosphere Reserve suffered from very high rates of deforestation in the 1990s. However, one of the few islands of forest cover maintenance that existed in the region in this period is a community with a well-functioning CFE (Merino Pérez and Hernández Apolinar 2004). Deforestation rates in many parts in Oaxaca have been high in recent decades (Velázquez et al. 2003), but the pine-oak forests of the Sierra Norte region, dominated by many highly diversified CFEs, has had 3.3 percent expansion of forest cover in these forests over a twenty-year period (Gómez-Mendoza et al. 2006). Thus, the available evidence suggests that community forest management has made a contribution to reduction of deforestation in both temperate and tropical areas and can be as effective or more effective than public protected areas in doing so. Agricultural abandonment has also led to forest recovery in many CFE communities, and in many cases there is no land use change permitted in the recovering forest; it is incorporated into forest management or ecotourism (Bray 2016; Robson and Berkes 2011; Robson and Klooster 2018). Of relevance, indigenous peoples, whether in ejidos or comunidades, have been estimated to occupy 20.2 million hectares, of which some 18 million hectares are covered by vegetation, including many of the forests of high conservation value, corresponding to some 10.3 percent of the national territory. Around half of the rainforests and tropical montane cloud forests, and a quarter of the temperate forests have been estimated to be in indigenous agrarian communities, almost always having low or nonexistent deforestation (Boege 2008). In sum, deforestation threats from traditional sources such as agriculture and cattle appear to be extremely minor in Mexican community forests, though emerging threats from forest pests such as bark beetles (*Dendoctronus spp.*) and mistletoe (*Psittacanthus calyculatus*) are becoming a more serious problem (Durán and Poloni 2014).

SILVICULTURE

A frequently expressed concern about forests managed for timber, even those managed more sustainably, is that even when forest cover and some functionality are maintained, biodiversity, structural complexity, habitat quality, and ecological resilience may be compromised. Although concerns continue to be expressed about ancillary impacts (Gibson et al. 2011, 380) and although logging impact can vary considerably depending on silvicultural practices in temperate and tropical forests, studies nonetheless have shown that selectively logged forests can retain high indices of forest taxa and have relatively benign impacts on biodiversity.

TEMPERATE SILVICULTURE

CFEs have likely contributed to reduced deforestation, to forest maintenance, and to forest recovery and with relatively low impacts on biodiversity, despite the fact that they cut down trees. Their positive stewardship involves cutting down trees only in the zoned production areas of their territory and using government-regulated silvicultural practices that tend to maintain forest cover in a shifting mosaic of logged forests in various stages of succession. Silviculture can be considered applied forest ecology, managed anthropogenic disturbances of the forest. The natural dynamics of forest ecology are manipulated through methods such as logging intensity, thinning, and pruning that influence forest structure and species composition, responding to varying management objectives such as timber, water, wildlife habitat, or agroforestry (Smith et al. 1997). In the early period of logging in Mexico little applied forest ecology took place. In the temperate areas in the center and north of the country, the first major commercial harvests of the modern period in the late nineteenth century, were clear-cuts under concessions along railroad lines. Destructive industrial logging in a concession given to the U.S.-owned Sierra Madre Land and Lumber Company began in Chihuahua in 1906, and the 1911 onset of the Mexican Revolution led to further forest devastation for fortifications, railroad ties, and firewood (Boyer 2015). In the Lacandón rainforest, what J. de Vos (1996) called the golden age of mahogany exploitation occurred from 1895 to 1913, conducted by U.S. logging companies with Mexican partners under a forest concession model. The environmental impact of this period was relatively minor; extraction stayed limited to mahogany and cedar and focused only along the rivers. Therefore, the first modern emerging SES of Mexican forestry was based on concessions to foreigners, punctuated by disruptions of the revolution, with relatively limited impacts in what were still vast, thinly populated forests.

The 1926 Forest Law (see chapter 2), the first national effort to regulate the disordered forest extraction of the former period, had little initial impact, but among other things it called for management programs with silvicultural prescriptions and the banning of clear-cutting (Boyer 2015; Torres Rojo, Moreno-Sánchez, and Mendoza-Briseño 2016). The modestly more regulated logging through the forest cooperatives (see chapter 2) in the 1930s followed the restriction against clear-cutting, conducting selective logging practices, taking out desired specimens of a particular species, the norm in Mexican forestry for decades. The SES of Mexican forestry now manifested itself in impacts on the structure and composition of the forest through efforts to make silvicultural practices more uniform. Selective practices were intended to maintain a continuous upper canopy composed of healthier, larger trees that could keep growing, but in practice, loggers targeted the largest and best-formed trees, leading to a decline in forest quality and changes in structure (Torres-Rojo,

Moreno-Sánchez, and Mendoza-Briseño). The next stage of regulation of silvicultural practices, intended to control these earlier practices, occurred with the development in 1944 of the silvicultural standard known as MMOM. This occurred in the context of the UIEFs and the 1942 Forest Law (see chapter 2). MMOM was finally required nationwide in the 1964 Forest Law (Armendáriz Payán 2014), the first time that public policy left a clear standardized imprint on the managed forests of Mexico. MMOM is a selective method that harvests marked trees with a cutting cycle from seventeen to thirty-five years and with a diameter minimum. The practice sought a sustained yield, where estimated harvest volumes, obtained in samples in forest inventories, should not exceed the forest growth rate. It aspired to achieve a steady flow of timber by concentrating harvests in a single or limited number of annual cutting areas, and it involved planning of harvests of territories of up to 200,000 hectares, reflecting the realities of the concessions of the 1940s–1980s (Torres Rojo, Moreno-Sánchez, and Mendoza-Briseño 2016). These methods would later have to be scaled down to the management realities of individual community territories.

In 1980, the community forestry era in full launch mode, MMOM was further standardized and the name formally changed to the Mexican Method for Ordering Irregular Forests (Método Mexicano de Ordenación de Bosques Irregulares; MMOBI). MMOBI became standardized for the uneven-aged stands, stands with a variety of age classes and diameters, now typical of Mexican forests. MMOBI incorporates the use of the Liocourt Curve, which establishes the number of trees in different diameter categories, and manages toward the achievement of an "ideal" uneven-aged structure (Hernández-Díaz et al. 2008); it also continued to use exclusively selection cuts. The method has an environmentally friendly reputation, since it maintains a continuous canopy and more of the structure, if not the composition, of an uneven-aged natural forest. However, the use of selection cuts, which left only small gaps in the forest not conducive to the regeneration of sun-loving pine, was observed to encourage an ecological succession from pines to lesser-value oak and forest degradation (Snook 1986). For example, due to poor silvicultural practices in the northeastern Sierra Madre in Chihuahua, once healthy stands of pine mixed with oak became dominated by oak and shrub with a few, poorly formed pines left (Gingrich 1993). These observations and the ongoing scaling down of silvicultural practices to community territories of widely varying sizes and forest conditions, and the challenges of managing an uneven-aged forest, led to the development of the MDS in the early 1980s.

MDS (also called the seed-tree method) is an even-aged system with three to four intermediate thinnings distributed evenly over a rotation period of fifty to eighty years. The sequence of cutting typical of MDS methods consists of a varying series of precommercial thinning and thinning stages (*preaclareos* and *aclareos*), a regeneration cut that leaves seed trees (*árboles padres*), and a liberation cut that takes out the seed trees. The goal is to maintain a more even-aged forest, or to convert an uneven-aged forest to an even-aged one through a complete cutting cycle. The method leads to higher timber productivity due to the fast growth rate of the more open second-growth forests and higher timber yields from the thinnings. Management of silvicultural systems can be highly complex, and the thinning process can be adapted to maintain uneven-aged stands (Torres-Rojo, Moreno-Sánchez, and Mendoza-Briseño 2016). MDS leads to forests that are more intensively managed and, after regeneration cuts, leaves small (1–2 hectares) openings in the forest where pine regeneration can more easily take place. Natural regeneration has been observed to be vigorous in many Mexican forests. In the 1986 Forest Law, both MMOBI and MDS were required to include more sustainable forest management practices and multiple-use objectives, including provisions for wildlife habitat and biodiversity exclusion areas, spatial arrangements of specific stands, conservation corridors, and riparian buffer zones (Torres-Rojo, Moreno-Sánchez, and Mendoza-Briseño 2016). Finally, Finnish technical cooperation in the 1990s introduced a form of MDS called the Conservation and Silvicultural Development System (Sistema de Conservación y Desarrollo Silvícola; SICODESI), which incorporated software that modeled growth rates under different management scenarios (Hinojosa-Flores, Skutsch, and Mustalahti 2016).

The application of MDS has been highly uneven. It was not introduced in the Sierra Norte of Oaxaca until the early 1990s, and due to a new CONAFOR-SEMARNAT policy, MDS only began to be applied in Chihuahua for the first time in the 2012–13 harvest season. One danger of the larger openings left by MDS is that it can be vulnerable to land use change. One earlier attempt at implementing MDS in Chihuahua was stopped because community members were found to be planting potatoes in the clearings. In the 2000s even more intensive silviculture practices known as group selection and strip clear-cutting, taking out all trees in blocks or strips from 0.25 hectare to 1.5 hectares, were developed in the Sierra Norte of Oaxaca by Ixtlán de Juárez and are now being used by several other communities there. In both cases, natural regeneration is normally used, relying on the seed rain from the surrounding forest. The aesthetics of the small clear-cuts have created some controversy in the community, since the relatively large openings were quite visible from the road, but it has been accepted since natural regeneration has been so vigorous.

As Mexican foresters gained experience with MDS, they observed that a given forest could present diverse conditions, so they began combining MMOBI and MDS and variations in the same forest (Hernández-Díaz et al. 2008) or using "combined methods." Combined methods are commonly used in forests of 2,000 to 12,000 hectares depending on soil type, aspect, slope, biodiversity, forest health, and timber stock (Torres-Rojo, Moreno-Sánchez, and Mendoza-Briseño 2016). San Pedro Nuxiño, in the Mixteca region of Oaxaca, is an example of a combined methods approach, and it is also typical of many Type II communities with small forests (Bray and Duran-Medina 2014). San Pedro has a total of 6,311 hectares of territory and about 3,000 hectares of forest. Of this forest, around 2,000 hectares has been informally privately parceled where unauthorized logging occurs for firewood and charcoal. The remaining 1,000 hectares are managed as a common property by the community CFE. The forest was logged under concessions from the early 1970s until 1989. During this period the community had little knowledge or participation, logging roads were built in a haphazard fashion, and a form of MMOBI was practiced that took out only large pines. These practices drove a transition to a significantly oak-dominated forest. The community stopped all logging in the forest in 1989 due to concerns about overexploitation, and from 1989–2004 it was protected by the community. About half of the forest was impacted by a fire in 1991, and after that the community initiated some reforestation activities. Supported by a forester, the community got an approved management program and began logging for the first time as a CFE in 2005. However, as a result of the previous silvicultural practices, the forest was estimated to be about 70 percent oak and 30 percent pine. Due to the varying conditions of the forest, and with goals of both logging the oak for charcoal and restoration to a pine-oak forest, both MMOBI and MDS are used in different parts of the forest, the latter used to clear out oak-dominated stands and to open sufficiently large spaces to encourage pine regeneration. As well, the community is slowly rationalizing the placement of logging roads, eliminating the more haphazard placement from the concession period, and reducing the amount of forest dedicated to roads (Bray and Duran-Medina 2014).

Mexican forest and environmental laws call for a substantial amount of environmental protection in logging. The management programs have some similarities with RIL in Indonesia and elsewhere (Griscom and Cortez 2013; Putz et al. 2008). However, the Mexican management programs are focused on what trees are logged (the inventory), forest maintenance, and reduction of environmental impacts, whereas RIL is primarily focused on harvest practices (J. M. Torres Rojo, personal communication, April 25, 2018). B. W. Griscom and R. Cortez (2013) divide the RIL practices into categories of better harvesting, protection, and growth with specific activities under each one. Although not focused on the harvest, the elaborate Mexican regulatory framework does coincide in some areas with RIL (Bray and Duran-Medina 2014). For example, the RIL regime requires detailed inventories that map, mark, and measure the trees to be harvested (Putz et al. 2008). Such forest inventories are a basic requirement in Mexican regulations. RIL practices call for road and skid trail planning, which are also required in the Mexican regulations, and environmental mitigation regulations include directional felling. Protective practices such as riparian buffer zones and not logging in high conservation value forests are also called for in RIL and required in Mexican regulations. Because logging in Mexican temperate forests is frequently carried out on steep slopes, the leaving of slash in contours to control erosion is required and commonly practiced. Finally, silvicultural practices that ensure regeneration of native trees and long-term timber production, income, and employment are criteria of RIL, and these practices are also present in both Mexican regulations and community practices and norms (Bray and Duran-Medina 2014). Conservation in logging practices is evident in the Sierra Norte of Oaxaca, where it is common for communities to harvest less than the authorized amount for conservation reasons; ten of nineteen communities in one sample logged less than the authorized volume, for conservation purposes (Pazos-Almada and Bray 2018). On the other hand, in a study in Michoacán and Durango most communities approached 100 percent of authorized volume. In the cases where communities harvested less than authorized, the reasons given were lack of demand or lack of commercial timber, not conservation (Antinori and Rausser 2010).

TROPICAL SILVICULTURE

The most important geographic setting for tropical silviculture has been the central and southern part of what is today the state of Quintana Roo and to a lesser extent the state of Campeche, both in the Yucatán Peninsula. Virtually uninhabited for centuries after the collapse of the Ancient Maya civilization, the area was resettled by Maya refugees fleeing religious taxes and expanding henequen plantations in Yucatán State in the mid-nineteenth century. These "Santa Cruz" Mayans mounted an armed uprising, known as the Caste War (1847–1901), with an estimated 50,000 refugees from their eventual defeat initially populating the forests but declining to some 12,000 by 1874 (Bray and Klepeis 2005; Konrad 1991). The Mayans were briefly brought under the control of the state in the early twentieth century but were largely abandoned to the forest again until the Cárdenas presidency in 1934. Some of the early agrarian reform ejidos were established in Quintana Roo in the 1930s as "forest reserves" to encourage chicle production,[8] the sap of the chicozapote (*Manilkara zapota*) tree used in the production of chewing gum. With 420 hectares per person calculated to allow for a continuous harvest of resin, ten ejidos established from 1935 to 1942 averaged almost 35,000 hectares each (Bray and Klepeis 2005; Hostettler 1996), assuring the conservation of vast forest areas. In the first half of the twentieth century, mahogany and Spanish cedar (known as *preciosas*, or precious timbers) were logged by unregulated small contractors. The forests of central and southern Quintana Roo were not placed under systematic management for timber until the granting of a 550,000-hectare concession to MIQROO in 1956. MIQROO developed the first FMP in tropical America (Flachsenberg and Galletti 1999). It established a tropical version of the MMOM that included forest inventories and selection cuts in twenty-five-year cycles, on the assumption that it took seventy-five years for a mahogany tree to reach full maturity, thus termed a polycyclical system. Annual harvest volumes were determined for the eight most valuable species, principal among them mahogany (*Swietenia macrophylla* King) and Spanish cedar (*Cedrela odorata*), each most desirable at a minimum diameter of 60 centimeters. Large numbers of LKS were ignored, since they had uncertain markets. The valuable commercial timbers occurred in the forest at very low densities, only around 2 percent of the biomass, making selection cuts the only practical practice. Silvicultural practices made no effort to improve the structure, species composition, or density of the forest, and, as will be shown below, harvest volumes fluctuated considerably from year to year. However, commercial trees were carefully marked and roads constructed to

---

8. Nearly all of the ejidos in central Quintana Roo are populated by Mayan indigenous peoples. But they were not given land grants as indigenous comunidades because they did not have titles given by the Spanish Crown in the colonial period.

reach only them, reducing environmental impacts (Snook 1998; Torres Rojo, Moreno-Sánchez, and Mendoza-Briseño 2016).

In 1983, due to concerns about deforestation in Quintana Roo because of tourism development, the PPF was established as an innovative multilevel stakeholder partnership between the Mexican government, through the DGDF, the German development agency GTZ, and the Quintana Roo government (see chapter 2). The PPF never had any formal legal status and was consequently a "linking pin" organization, an organization that has no formal authority but is able to mobilize coalitions around a specific issue (Carley and Christie 2000; Jönsson 1986), in this case more sustainable management of Quintana Roo's tropical forests. New inventories were also conducted using participatory methods. As a major step, each community defined PFAs, where there could be no agriculture or land use change, only controlled extraction such as logging, chicle tapping, hunting, and collection of polewood and palm thatch for construction. Mahogany, like pine, requires larger gaps in the forest for successful regeneration, historically created by hurricanes, subsequent fires, and, most notably, milpa agriculture. As a result of these natural and anthropogenic disturbances, mahogany typically occurs in discrete, essentially even-aged patches across the landscape (Snook 1998). To harvest the trees, the PPF used the same basic silvicultural practices as MIQROO, attempting to assure sustained yields of mahogany from the forest by placing controls on the spatial distribution and rate of harvesting, using the polycyclical system and a diameter limit of 55 to 60 centimeters. By 1992 the basic silvicultural model had been expanded to four associations of forestry ejidos that had a combined allowable cut of 10,580 cubic meters per year of mahogany and cedar from 393,481 hectares of PFAs (Flachsenberg and Galletti 1999).

In the first twenty-five-year cutting cycle, mahogany trees larger than the diameter limit are harvested, which in practice was an average of one stem per hectare. This included many large mature mahoganies left in previous harvests because they did not meet MIQROO requirements. L. K. Snook (1998) argued that these practices altered the size and age structure of the forest with likely negative impacts for wildlife and biodiversity. In year 26, the beginning of the second cutting cycle, the trees formerly in the classes of 35 to 54 centimeters, now presumably above 60 centimeters, will be harvested, and in year 50, the third cycle, trees in the smallest size classes at the beginning of the cycle will be harvested. However, there is evidence that 90 percent of the mahogany trees are growing more slowly than the projected rate and that both minimum diameters and total volumes of harvests in the second and third cutting cycles would be considerably lower, leading to an impoverishment of the forest of large mahogany trees. As well, the selective logging practices do not create large enough spaces for the shade-intolerant mahogany seedlings. The traditional openings left by shifting agriculture were ideal for regeneration, and excluding their use from the PFAs has been argued to be an error (Snook 1998). Quintana Roo foresters have proposed more intensive silvicultural practices that would open canopy gaps (bosquetes), but regulatory rigidity and unclear financial feasibility have hindered its adoption (E. A. Ellis et al. 2015).

The results of some eighty years of shifting silvicultural practices for the harvests of mahogany and Spanish cedar are shown in figure 16.

The historical impacts of shifting silvicultural practices for mahogany and cedar represented in figure 16 can be analyzed in three historical periods (Bray 2004). From 1938 to 1956, logging was carried out by small private concessionaires in an uncontrolled fashion with wildly fluctuating annual volumes between 10,000 and 50,000 cubic meters. The sharp spike in 1956 results from salvage logging after the devastating Hurricane Janet in 1955. From 1956 to 1983, logging was carried out under the concession to MIQROO with better management practices, as noted above, but with only a moderate reduction in annual volume fluctuations, generally between 30,000 and 50,000 cubic meters.

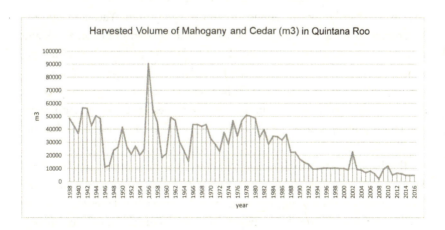

FIGURE 16. Harvested volume of mahogany and cedar (cubic meters) in Quintana Roo (1938–2016). Data courtesy of Dachary and Arnaiz B. (1983); Informe de labores del gobierno de Pedro Joaquín Coldwell, 1987–1990. Análisis Estadístico de los Estados Unidos Mexicanos, ediciones 1990 and 1988–1989, INEGI (1992 and 1990 and 2002–16).

Beginning in 1984, community management produced a striking reduction and stabilization of harvests in three stages. In the first stage, from 1984 to 1988, harvests were 22 percent lower than in the previous five-year period (1979–83) with MIQROO. After this initial period of logging, the PPF foresters realized there had been problems with measurements in the participatory inventory, and they carried out a new inventory resulting in more reductions in the harvests, in the second stage, from 1988 to 1992. As of 1993, in the third stage, volumes were further reduced over concerns of overharvesting, where they remained stable until 2005 at around 10,000 cubic meters. For the 1993–2005 period, harvests represented a 78 percent reduction from the last five years of MIQROO. R. E. Rice et al. (2001, 16) argued that "most logging companies in the tropics engage in the rapid harvest of a limited number of valuable tree species because it is profitable." While this was true of the small loggers and MIQROO, it was manifestly not the case with the CFEs. However, after 2005 harvest problems became more manifest. The sharp drop in 2008 is due to Hurricane Dean and the sharp rise afterward due to post-Dean salvage logging, but from 2011 to 2016 harvest volumes of mahogany and cedar had dropped to an average of around 5,000 cubic meters, half of what they were from 1993 to 2005. By 2011, many of the communities had started on the second twenty-five-year cycle in the poly-cyclical system, and, as Snook (1998) predicted, harvest volumes have dropped precipitously from already low levels. However, it has also been pointed out that the current distribution and abundance of mahogany in the Yucatán Peninsula is superior to stocks in Peru and Bolivia (Navarro-Martínez et al. 2018), and foresters report that reserves are good and foresee no problem in maintaining production at least at the current reduced levels. Declines in production are also not necessarily related to forest stocks but may also be due to improved markets for some LKSs and to forest communities not harvesting the entire authorized volumes for organizational reasons (E. Ellis, personal communication, June 12, 2019).

This harvest history makes clear that one of the most important challenges faced by CFEs in tropical Mexico is their historical legacy of forests impacted by decades of overharvesting, leaving them with impoverished forests presenting challenges for sustainable forest-based livelihoods. Due to the reduced income from the precious timbers, the harvests have shifted decisively toward the LKS, constituting around 70 percent of the total harvest over the past two decades. However, most of these trees are logged at well below authorized volumes. As a further indication of the high degree of conservation associated with logging in tropical Mexico, the silvicultural practices leave large areas of intact, conserved forest. A study that placed Mexican selective logging practices in a global comparative perspective found that in a comparison of six tropical countries, Mexico had the highest percentage of intact forest remaining in logged blocks, from 77–97 percent intact (Putz et al. 2019).

For the 2012–13 season, mahogany was harvested at 72.6 percent of the authorized volume, but the average for all species, including a large number of lesser-known tropical species, was only 30.1 percent (E. A. Ellis et al. 2015). LKS used for polewood are the most closely studied group (Racelis and Barsimantov 2008; Sierra-Huelsz et al. 2017). Polewood was traditionally used by the Mayans for housing, but the booming Cancún-Tulum tourism corridor created a strong market demand. Some thirty-two species of tropical hardwoods are harvested for polewood from the forest understory, not secondary succession. Concerns about overharvesting have been expressed, chief among them likely reductions in forest structural diversity, species composition, and possibilities for regeneration. An increasing number of high-quality polewood species are seen as disappearing from the forest, and polewood constituted up to 10 percent of the total timber harvest in a sample of twenty-two communities. The government permitting process, organized around large commercial species such as mahogany, is not well adapted to the species diversity and other characteristics of polewood, which threaten to turn it into an open-access resource (Racelis and Barsimantov 2008). "Conflict of use" issues have also arisen alongside steps taken to ban the use of traditional timber species as polewood. For example, an emergent demand for posts from the chicozapote tree (*Manilkara zapota*), traditionally used for chicle tapping, may endanger that traditional product (Sierra-Huelsz et al. 2017).

Recent studies of silviculture globally and in Quintana Roo and Campeche (Yucatán Peninsula) have focused on how RIL can reduce forest disturbance and carbon emissions (P. Ellis et al. 2019; E. A. Ellis et al. 2019). The Yucatán Peninsula study of a sample of ten communities found that silvicultural practices and adherence to RIL practices, explicitly or not, varied greatly in the sample. However, the per-hectare total carbon emission baseline was up to three times lower than similar studies in Brazil and Indonesia (linked to logging of much lower intensity in Mexico). With the measure of mean carbon emissions per volume of harvested timber, it was found that Mexican emissions were about midway among values for forests across the tropics. More widely and consciously applied RIL practices, such as directional felling and skid trail planning, by the communities could increase timber volumes and logging intensities while simultaneously reducing carbon emissions (E. A. Ellis et al. 2019).

CERTIFICATION AND FOREST MANAGEMENT

Environmental certification of forest products is an example of forest governance by market mechanisms or nonstate authority (Cashore, Auld, and Newsom 2004). It involves third party audits by an independent entity that is neither a producer nor a consumer, and usually it involves NGOs but can also be conducted by governments. The process uses a set of established environmental and social standards and grants a label indicating that certified products come from forests where management practices meet the standards (Klooster 2006). Mexico has two certification standards that provide partial governance of the SES of Mexican community forestry: a nonstate forest certification system operated by the Forest Stewardship Council (FSC) and the state-sponsored Mexican Forest Certification System (MFCS). Both systems have the goal of "measuring, monitoring, and enhancing sustainable forest management" (García-Montiel et al. 2017, 290). Mexico stands out globally because though only 3 percent of FSC-certified forests are communal, 35 percent of this total are Mexican forests (Gerez-Fernández and Alatorre-Guzmán 2007).

Established in 1993 by 130 stakeholders from producers, NGOs, foundations, and large retailers, the FSC assumed the goal of promoting environmentally appropriate, socially beneficial, and economically viable management of the world's forests. Environmental concerns include reduction of the environmental impact of logging and logging roads, the establishment of conservation areas, written management plans, elimination of toxic fungicides from sawmills and plans to deal with the piles of sawdust that accumulate near sawmills, among many others. Some of the first certified forests in the world began in Mexico with the certification of Ejido Caobas in Quintana Roo in 1993. As of 1998 the PROCYMAF program (see chapter 3) began promoting and subsiding certification in Oaxaca; later in other states and in the state government of Durango in the early 2000s, PROCYMAF promoted certification in meetings with forest communities and sawmills and subsidized certification (Klooster 2006; Taylor 2005).

Klooster (2006), writing before the Mexican system was established, suggested that Mexican CFEs have two barriers to certification: the barrier of getting certification in the first place, due to high costs and paperwork demands, and the barrier of using certification to reach markets. Big retailers dominate the certified commodity network, and demands for high volumes and low prices are difficult for CFEs to meet. Thus, the price premium from certification is insufficient to cover the costs of certification and the required improvements in management. As a result, as of the mid-2000s, only the largest CFEs in Mexico had gotten certified, and they generally did it as a seal of good forest management rather than for any direct economic benefit. However, studies have suggested that there are economic benefits to certification, despite perceptions to the contrary. For example, S. Humphries (2010) found in a sample of twenty-five ejidos in Quintana Roo that over time, certification contributed to higher prices for CFEs in QR and that the pricing system differentiated according to species, quality, and market, as well as attributes of sales and producers. M. Karmann and A. Smith (2009) also found in multiple countries, including Mexico, that FSC certification brought access to higher prices and better markets.

In the twenty-year history of certification in Mexico, there have been ups and downs in the number of certified community forests. However, as of June 2018, the number of certified forests by FSC was likely at an all-time high; sixty-four communities had certified a total of 1,122,946 hectares, thanks to significant government subsidies. Nearly half of this forestland (48 percent) is in Durango, suggesting that the early twentieth-century push by the state's government, now strengthened by federal support, has maintained itself. Another 32 percent of the forests are in Chihuahua, for a total of 80 percent in the two most important forestry states. Dividing this figure by the 6.2 million hectare figure from chapter 3 for the total number of forests under community management suggests that 18 percent of the total forestland is certified. While a substantial number, major potential clearly remains for a substantially higher number to become certified. The concern of Klooster (2006) about certification of only large forest communities appears to be have been allayed somewhat by the government programs. Twenty-three of the sixty-four certified communities in 2018 have fewer than 1,000 hectares. Puebla has a particularly high number of small certified forests; twelve of seventeen have fewer than 1,000 hectares and one, Ejido Jonuco Pedernales, has only 336 hectares. The smallest certified forest in Mexico is Ejido San Miguel Coapa in Michoacán, which occupies 264 hectares (from the FSC certificate database: "Welcome to the FSC Public Search" 2020).

Is there any evidence that FSC certification has environmental benefits? Blackman, Raimondi, and Cubbage (2017) carried out a study of 1,162 corrective action requests (CARs) for thirty-five FSC certified forests in Mexico. CARs are the requested changes in processes or forest practices and conditions that forest managers need to make in order to obtain or retain certification. The researchers found that most CARs did not focus on environmental issues but rather social and legal ones. The majority of CARs that focused on forest management and environmental outcomes were minor in nature, required only small changes, and were promptly addressed. Blackman et al. suggested that certification may produce a modest additional effect on environmental outcomes but that the minor nature of most environmental CARs suggest forests are already well managed.

The MFCS began in 2011 and was conceived as a more accessible alternative to the rigor and cost of the FSC. As of 2016 both systems had around 900,000 hectares under certification (as noted above, by 2018 the FSC had expanded to over 1.1 million hectares). García-Montiel et al. (2017, 15) conclude that the MFCS helps gain prestige in national markets and is sufficiently rigorous that it could be helpful in improving forest management and could serve in a "stepwise" fashion as a training ground for eventual FSC certification, since FSC does "provide the strongest seal for all international markets."

PAYMENT FOR ENVIRONMENTAL SERVICES AND VOLUNTARY CONSERVATION PROGRAMS

PAYMENT FOR ENVIRONMENTAL SERVICES

Mexico has had one of the world's largest PES programs implemented since 2003; it came to be primarily focused on Payment for Hydrological Services (PHS) (Muñoz-Piña et al. 2008; Alix-Garcia et al. 2009). With the express purpose of giving economic incentives to landowners, principally communities, to not deforest in areas that were both of hydrological significance and under threat from deforestation, it took on the additional goal of maintaining rural incomes and reducing poverty (Sims et al. 2014). Over the years of implementation, the program became increasingly focused on poverty alleviation and less on market efficiency and conservation due to the influence of civil society in the policy process (McAfee and Shapiro 2010; Shapiro-Garza 2013). The PES is paid to participating landowners directly by the state, which receives the money from funds earmarked from fees collected from water users. By 2010, the Mexican government reported that 4,646 communities covering nearly 2.8 million hectares were under contract (CONAFOR 2010), the majority on common property lands.

The PES in Mexico is administered by CONAFOR and is therefore a government-run program rather than a free-market relationship between service users and service providers. This setup fits the definition of R. Muradian et al. (2010, 1205) of environmental services payment as "a transfer of resources between social actors which aims to create incentives to align individual and/ or collective land use decisions with the social interest in the management of natural resources" and not the more strictly market-based definition by S. Wunder (2005). Studies have found that the PES effort was notably successful and that it "significantly reduced deforestation rates compared to what would have happened otherwise" (Alix-Garcia, Shapiro, and Sims 2012, 633). K. R. Sims et al. (2014) found that the PES program had successfully implemented adaptive management, without explicitly using the concept, and that the program had met its criteria by enrolling lands that were more commonly located within overexploited aquifers, priority mountain areas, and protected natural areas. As well, the enrolled areas had lower mean surface water areas than all forested lands, were closer to population centers of over 5,000 inhabitants (where demand for hydrological services is greater), and had more cloud forest (which had been given priority) than all forest areas. A recent study also found that the PES program reduced both forest cover loss and forest fragmentation both regionally and nationally (Ramirez-Reyes et al. 2018). CFE communities were eligible for the PES program, though areas under contract could not overlap with the areas managed for timber, allowing CFE communities to obtain income from areas that they were already conserving. Kerr, Vardhan, and Jindal (2012) found in a choice experiment in five villages in the PES program that the community with the highest participation in the PES program was also the only one with an FMP, suggesting the collective action around forest management supported collective action around the PES program.

J. M. Alix-Garcia, K. R. Sims, and P. Yañez-Pagans (2015) studied the impact of the PES program in poverty alleviation and conservation throughout a quasi-experiment (treated and control groups evaluated at two different periods). The impact on poverty was evaluated at household level with a total sample of 1,210 households (private and community members). The researchers found a marginal impact on alleviating poverty and concluded that "PES is more of a win-neutral than a win-win strategy for environment and development" (Alix-Garcia, Sims, and Yañez-Pagans 2015, 32). They attributed the low impact of the program in alleviating poverty to the high participation costs due to increased forest management and maintenance by beneficiaries.

Evidence indicates that the interaction between the OTC, ADCV, and PES programs created new incentives for collective action around forest conservation and consolidated a culture of conservation in forest communities, including in those without logging permits. Two outstanding examples of communities, both in Oaxaca, have organized themselves around protection and biodiversity conservation after participating in OTC processes and with incentives provided both by the ADCV program and PES. The first is a six-community organization, the Natural Resource Committee of the Upper Chinantla (Comité de Recursos Naturales de la Chinantla Alta; CORENCHI), in the Sierra Norte of Oaxaca. These communities, supported by the NGO Geoconservación, are protecting cloud forest and montane tropical forest with high degrees of biodiversity (Meave et al. 2017). CORENCHI is an example of collective action at multiple levels, community and intercommunity organizational processes that have established contiguous ADCVs based on strict conservation on 77 percent of the committee's 34,908 hectares of territory (Velasco Murguía, Durán Medina, and Bray 2014). This highly turbulent process, which required integrating rules from multiple levels and extensive negotiations of conflicts in multiple arenas, depended on community leaders who had gained experience in community collective action through a coffee cooperative and learned how outsiders valued their forests for their floristic diversity through working with university researchers. This information gave community leaders a new appreciation for the conservation value of their forests and for the idea that conservation might be used to generate income to reduce emigration. The implementation of OTCs funded by PROCYMAF, the development of community statutes, the establishment of ADCVs in all six communities, and the income from the PES program—all government or multilateral programs—were key incentives in collective action around strict conservation (Anta Fonseca and Mondragón Galicia 2006; Bray, Durán, and Molina-González 2012). This suite of conservation-oriented programs showed in the case of CORENCHI that social capital can accumulate even in a context that had already strong prosocial norms, can encourage regional collective action, can display a perception of equity in benefits distribution, and can achieve a degree of poverty alleviation in community forests managed for conservation, not for timber. A study of the impact of the PES in two of the CORENCHI communities concluded that "government policy has strengthened community governance processes that reinforce and make more conscious and active historical virtuous behavior toward forest management" (Nieratka, Bray, and Mozumder 2015, 353–54).

The second conservation-oriented intercommunity organization is the Community System for the Management and Protection of Biodiversity (Sistema Comunitaria para el Manejo y Protección de la Biodiversidad; SICOBI), which had the support of the NGO the Autonomous Group for Environmental Research (Grupo Autónoma para la Investigación Ambiental; GAIA). SICOBI, a nine-community organization that started with one community in 1998 and expanded beginning in 2001, used the OTC methodology to analyze the physical, biological, and cultural aspects of community territories to carry out land use zoning and planning and the processes and rules that could implement the vision resulting from the process (González and Miranda 2003). SICOBI has developed coffee agroforestry, honey production, bean cultivation, and mezcal production, and with ADCVs and the PHS program the organization is conserving 55,058 hectares of pine-oak, montane tropical, and dry tropical forests, some 56 percent of its total territory (Robson 2007; Sánchez Jacinto 2011).

These experiences are notable because they show collective action on the commons around conservation through extensive networks constituting multilevel governance. In these cases, the resource has value not for harvests but for strict conservation and generation of environmental services. These examples show that intercommunity organizations can establish rules that govern land use in the individual communities, that visionary leaders can overcome traditional animosities between communities, and that conflicts are not "adjudicated" but negotiated in complex, turbulent settings at multiple scales (Bray, Durán, and O. Molina-González; Ostrom 1990). In short they shine a light on "community-based management, less of natural resources, than of extensive networks of government and civil society actors" (Bray, Durán and O. Molina-González 2012, 154).

The OTC program in particular had a significant impact on forest policy. The emergence of the requirement of land use planning for the entire community territory that appeared in the 2003 Forest Law is due to the influence of the OTC methodology used in PROCYMAF since 1997 and first developed by the NGO ERA. There is evidence that in forest communities that were not logging, these programs served to greatly expand a consciousness of conservation, particularly with the PHS program providing clear economic incentives for conservation. For example, two small recently formed agricultural communities in the montane tropical cloud forest region of the Sierra Norte of Oaxaca who have received PHS payments, Santiago Cuasimulco and Nuevo Rosario Temextitlán, have high forest cover and informal and formal community rules against further deforestation despite the recommendation of their OTC that a modest expansion of agricultural lands could help meet community needs. This culture of conservation apparently emerged as a result of both the OTC planning process and the PES program (Van Vleet, Bray, and Durán 2016), though it could be vulnerable to eventual reductions in the PES program. These programs appear to have created an entirely new sector of community-managed forests in Mexico, those managed consciously primarily for conservation. This new development in Mexican forest policy is little noted and warrants further analysis and documentation.

VOLUNTARY CONSERVATION AREAS

A reform of Article 59 of the General Law of Ecological Equilibrium and Environmental Protection (Ley General del Equilibrio Ecológico y la Protección al Ambiente; LGEEPA) in 1996 allowed communities to set aside land for conservation with official recognition. However, it was not until 2003 that the National Commission of Natural Protected Areas (Comisión Nacional de Áreas Naturales Protegidas; CONANP) started a program of certifying communal and ejidal reserves, the ADCVs, a new federal protected area category designed to include community and private reserves. The ADCVs are incorporated into the National Registry of Protected Areas (Registro Nacional de Areas Naturales Protegidas) according to Article 74 (Martin et al. 2010). However, they are different from other public protected areas in several respects. The process of recognition is called "certification," is carried out after a specific request by communities, and is time limited, for periods of up to twenty-five years. Public protected areas, in contrast, require community consultations but are enacted by presidential decree and are in perpetuity. The communal conservation areas recognized under this system can be considered what the International Union for the Conservation of Nature calls "indigenous/community conserved areas" (I/CCA) (Borrini-Feyerabend, Kothari, and Oviedo 2004). Oaxaca has been a national leader in the establishment of ADCVs and for other forms of community conservation. Community protected areas in Oaxaca have been classified as community conserved areas (not formally recognized), ADCVs, conservation areas in forest management programs, sacred natural sites, and wildlife management areas. A 2011 survey found a total of 126 community conservation areas of all kinds in Oaxaca covering 375,457 hectares. This total was 14.5 percent more than the 327,977 hectares in presidentially decreed Protected Natural Areas in the state (Martin et al. 2011). This total of the conserved areas in forest management programs are not included, so this is almost certainly a serious undercount. As of 2019, nationally there were 216 communities with ADCVs, including 425,985 hectares. Of these, Oaxaca had 135 of the total 216 (62.5 percent) but only 29 percent of the total hectares, since there were some very large ADCVs elsewhere in Mexico (Comisión Nacional de Áreas Naturales Protegidas 2019). One study argues that both ADCVs and the PES program have impacted food production and sovereignty in communities in the Chinantla region of Oaxaca (Ibarra et al. 2011), though this conclusion has been challenged by Berget, Durán, and Bray (2015). However, the key takeaway here is that the OTCs, ADCVs, and PES programs were all substantial new government programs, had significant or exclusive foci on forest conservation and apparently contributed toward strengthening or creating a culture of conservation in forest communities, and had at least modest positive economic impacts.

BIODIVERSITY

As noted above, concerns are frequently expressed about the impacts of logging on biodiversity. Unfortunately, there has been little direct research on biodiversity in community forests managed for timber, but available evidence suggests that it is quite high. For example, in the tropical forests, G. Ceballos et al. (2005) found healthy populations of jaguars in CFE forests in southern Quintana Roo. M. W. Tobler et al. (2018) found in well-managed logging concessions in Guatemala and Peru that logging had no negative impact on more than twenty medium and large-sized mammals and that carnivores commonly used the logging roads as movement corridors. It is highly likely that similar results would be found in the community forests of Mexico. J. J. Figel, Durán, and Bray (2011) found evidence of a persistent population of jaguars in a region not considered ideal habitat for them and emerging attitudes in favor of jaguar conservation in the CORENCHI communities discussed above. A study in the Sierra Norte of Oaxaca found that the silvicultural practices there were compatible with maintaining the diversity and structure of medium and large mammals, due to the small size of the logging areas and large amounts of surrounding forest cover

(Hernández-Rodríguez et al. 2019). The impact of logging as practiced in Quintana Roo was found to be "benign" for bird species (Lynch and Whigham 1995). L. Ochoa-Ochoa et al. (2009) found that various forms of community protected areas, both formal and informal, were more effective than public protected areas for conservation of endemic amphibians. Almazán-Núñez, Puebla-Olivares, and Almazán-Juárez (2009) report relatively high counts of birds (117 species) in pine-oak community forests in Guerrero. As well, the implementation of locally enforced hunting bans in community forests has been widespread, since employment in CFEs or income from conservation has reduced the need for it (Bray et al. 2003; Figel, Durán, and Bray). Another contributing factor to high rates of biodiversity is that harvest intensities are generally low, and it has been found that species richness of invertebrates, amphibians, and mammals is higher in forests that have been selectively logged at lower intensities (Burivalova, Şekercioğlu, and Koh et al. 2014).

The Sierra Norte of Oaxaca is probably the most intensively studied community forest region for its biodiversity, though it is seldom mentioned that it is entirely a community-owned and -managed landscape (Pazos-Almada and Bray 2018). The Sierra Norte is the world's greatest center of endemism for the genera *Pinus* and *Quercus* (Gómez-Mendoza et al. 2006). K. Brandon et al. (2005) showed that Sierra Norte has among the highest biodiversity in birds, mammals, and amphibians in Mexico. For all of Oaxaca, 183 terrestrial mammal species have been recorded, Sierra Norte having the highest reported occurrence (Briones-Salas and Sánchez-Cordero 2004; Illoldi-Rangel et al. 2004), and new species are commonly being discovered (Briones-Salas et al. 2012; Canseco-Márquez, Ramírez-González, and Gonzalez-Bernal 2017). In the production forests the varying silvicultural practices, from selective logging to the small clear-cuts, create a structural mosaic that is likely to accumulate beta-diversity to maximize regional gamma-diversity (Socolar et al. 2016). Although some ecologists have called for greater protection measures in these forests (Brandon et al. 2005; Illoldi-Rangel 2008; Rojas-Soto, Sosa, and Ornelas 2012), it is apparent that the current high degree of biodiversity has been maintained in landscapes logged for over fifty years and managed by communities for over thirty years, with hundreds of years of human habitation and forest use before that (Van Vleet, Bray, and Durán 2016).

Currently, a fully formed Mexican community forestry SES has taken shape, with forest laws and policies influencing forest structure and composition in the millions of hectares of managed forests dispersed across the country. The ecology of the community-managed forests of Mexico has been shaped, in the first place, by successive versions of FMPs, supplemented by OTCs in the last two decades. Where community forests dominate the landscape, they have had a measurable impact on the reduction of deforestation and degradation. Varying forms of silviculture combined with the zoning practices have created community-managed economic spaces with a distinctive mosaic of forests managed for timber and for various forms of conservation. The relatively modest but growing percentage of certified forests have further standardized forest management at a regional and national scale. State conservation programs such as the PES initiative and ADCVs have complemented and given economic incentives to a new emerging culture of conservation to accompany the preexisting culture of industrial forestry. The combination of all of these elements has made important contributions to creating viable habitat for Mexico's abundant forest biodiversity.

CHAPTER 8

MEXICAN COMMUNITY FORESTS IN THE ANTHROPOCENE

Forest Resilience and Climate Change

The SES of Mexican community forestry has now been evolving for some one hundred years, beginning with the foundations for state-recognized community governance in the Mexican Constitution of 1917. Hundreds of CFEs have reached a state of maturity and consolidation that strongly suggest that they have the resilience to endure for decades to come even in the face of political, economic, and climate shocks. A number of others struggle with corruption, disorganization, and assaults from organized crime. However, those that have been successful for decades in managing their forests and generating incomes for their communities show the potential of the sector both nationally and globally as a model. The empirical success of this large-scale national experiment challenges the way we think about common property theory, resilience theory, and reforms to forest sectors globally. The Mexican experience also stands out in an international comparative framework for its accomplishments and lessons. However, it is crucial to ask if the adaptability and transformability of the SES will continue into the future. What are the practical prospects for further expansion of the sector nearly fifty years after the Echeverría reforms and community responses of the early 1970s launched it as the dominant aspect of the Mexican forest economy? Most important, and by way of closing out this analysis and narrative, can the sector help to mitigate climate change, adapt to climate change, and remain resilient? The planet and

Mexican community forestry are now plunging into the Anthropocene with unpredictable consequences. I will argue that the Mexican experience presents a model for how the SES of forested areas of developing countries can weather the political, economic, and meteorological storms of the future and can contribute to the challenging goal of holding global warming to 1.5 degrees Celsius above preindustrial levels (IPCC 2018). The Anthropocene, the naming of the current geological era on the basis of human impact in geological stratigraphy and earth systems in general, carries mostly negative connotations. However, the Future Earth project has developed a program entitled "Seeds of a Good Anthropocene" (E. C. Ellis 2018). The Mexican community forestry SES, a thoroughly anthropogenic sector, provides a model for a "Good Anthropocene" for developing country forest communities as an innovative strategy for connecting people and nature in sustainable ways. Mexican community forestry provides both theoretical and empirical lessons for a Good Anthropocene.

THE EVOLUTION OF MEXICAN COMMUNITY FORESTRY: THE THEORETICAL LESSONS

I have argued that it is state policy and markets interacting with emergent community collective action around common property forests that have constituted the political economy of reform in resilient Mexican forests. What follows is a synthesized political-economic interpretation of the data in chapters 2 and 3 that resulted in the institutional and organizational innovations discussed in chapters 4 and 5, the economic result of generally profitable CFEs that survive despite many competitiveness issues nationally and internationally and with generally positive ecological outcomes for Mexican forests, as argued in chapter 7. The dominant pattern of the Mexican political economy during most of the twentieth century was authoritarian. A long-enduring political party used clientelism to "divide and conquer," resting in part on authoritarian regional and rural elites and with substantial state direction of the economy (Fox 1992). Nonetheless, reforms in the forest sector in particular created a democratization of natural assets (Boyce and Shelley 2003) and a safe economic and political space where alternative and counterhegemonic practices could emerge (even in an authoritarian and later neoliberal context). The 1926 Forest Law was the first effort to provide state direction to the forest sector, but the foundations for Mexican community forestry were laid in the political economy of the Cárdenas period (1934–40). Cárdenas, in what has been called a "state-peasant alliance," called for agrarian communities to "become the central politico-economic actor in the countryside, supplying food and raw materials" (Fox 1992, 48), and one important raw material was timber. The state achieved a degree of autonomy from elite landed interests and instituted major structural changes in Mexican society (Fox 1992; N. Hamilton 1982). One of these little-recognized structural changes was the cooperatives of the Cárdenas period, the first stage of the carving out of a local political-economic and institutional space where community forestry entrepreneurial activity could take place and resilience could begin to emerge. The 866 forestry cooperatives organized by 1940 (Boyer 2015) were the first poorly implemented draft of reforms that would emerge again in the 1950s, on the basis of the revolutionary ideal that communities could manage their own forests. The top-down governance template established for ejidos and comunidades served as a crucial institutional and organizational framework that slowly and painfully encouraged community response over decades.

As noted in chapter 1, timber has been a commodity frontier linked to the global accumulation of capital (J. Williams 2007). Foreign capital was dominant in Chihuahua and Durango in the first half of the twentieth century, but most logging was done by small-scale logging companies (Boyer 2015). The lapsing of the community control ideal for most of the 1940–70 period of the counterreform was not absolute and bore within it the seeds of a transformed SES, its seeds coming from both above and below. As the agrarian reform continued in subsequent decades, pushed by communities demanding access to land and forests, it advanced into mountainous areas and the tropics; along the way, more and more of the natural capital of forests were redistributed, as in Durango and Chihuahua in the 1950s and 1960s and as in Guerrero and Chihuahua in the 1970s. These advances were politically possible because private forest capital was mostly relatively small and local and had reduced political power, creating the conditions for state power to frequently support land titling and community forestry. The efforts by INI in the 1950s in Chihuahua were the first state effort since the 1930s at creating new CFEs, and the CFEs were more focused and enduring than the cooperatives. In the 1960s, grassroots mobilizations in Durango against the asociaciones en participación, encouraged and supported by state reformers, were an example of the kinds of both top-down and bottom-up dynamic interactions that would characterize the sector in some regions (García-López 2013). Also, in Durango, relations with private logging companies were important in training and supplying human capital and in launching the first CFEs in Mexico in the late 1960s.

The second great populist reform period that more proximately helped to create the contemporary sector was the Echeverría presidency (1970–76). The Echeverría reforms constituted a "populist shift in official agrarian policy" and increasing state intervention in rural production (Fox 1992, 56), including timber production and new legislation that encouraged the formation of second-level or intercommunity ejido unions. Echeverría showed the political will to expropriate forest lands from both small regional logging companies, such as in Guerrero, and from politically powerful timber interests such as Bosques de Chihuahua, shifting control to parastatals, as reviewed in chapter 2. He also intervened directly in individual-state-level conflicts on the side of community forestry. This action foreclosed any further role for foreign capital in the forest sector and placed serious limitations on the role of national capital, a crucial element in opening up an insulated political-economic space for CFEs largely independent from private capital needs, due to both state subsidies and self-financing. The top-down reforms by FONAFE endowed hundreds of agrarian communities with human and physical capital through training and sawmills in the concessions areas. Later in the 1970s, the DGDF worked with small communities in areas that had been under logging bans; in the early 1980s the DGDF moved into more politically sensitive areas with concessions, such as Oaxaca and Quintana Roo, bestowing endowments of social, human, and physical capital. Also, in the 1970s, parastatals in Chihuahua, Durango, Guerrero, Quintana Roo, and elsewhere acted as regional forest development agents and began organizing and training communities to operate their own CFEs, even if still tying them to a single buyer. When in later years the parastatals became more corrupt and unresponsive, communities mobilized, commonly with federal government support, to demand further market openings for community forestry.

The analysis by Fox (1992) of food policy from 1980 to 1982 works well for the dynamics of forest policy as well. The top-down FONAFE reforms, the founding of the DGDF in 1974, and the early reformist period of the big parastatals reflected a "recurrent reformist current within the state" and the populist shift toward the economic empowerment of agrarian communities (Fox 1992, 40; Silva 1994), including forest communities. These reforms in the forest sector came almost entirely came from above: Durango in early periods was the only state where grassroots mobilizations played a role in shaping the sector, followed by Oaxaca in the early 1980s. The most significant grassroots uprising after the successive waves in Durango was Oaxaca with ODRENASIJ in the early 1980s. The reform forces within the government achieved the pinnacle of forest policy power with the ascension of the most important political entrepreneur of the period, León Jorge Castaños, as forest subsecretary in the early 1980s. By 1986 he pushed through the Forest Law that ended the concession period and gave CFEs legal legitimacy on several fronts. The presidency of Salinas de Gortari (1988–94) is virtually the only period from 1970 to the present in which the state showed indifference or hostility to the community forestry sector, focusing on plantation policy, but constitutional reforms still resulted in forest communities having a more complete bundle of property rights. New state forest programs from above emerged in the Zedillo presidency (1994–2000) in the PROCY-MAF and PRODEFOR programs, which pumped new social, human, financial, and physical capital into the sector. These programs were continued to varying degrees throughout the next three presidencies, through 2018, and the last three years of the Calderón presidency (2009–12) were a particular high-water mark for state efforts to support the sector, even if greater funds went to reforestation and environmental services in "passive conservation." The Peña Nieto presidency was a sort of holding period for the sector, in which support continued but with no significant new initiatives, with pressures on community forests to increase production, and with possible increases in downward mobility in vertical integration of the CFEs. It was in the 2000s that CFEs also began to experience heavier impacts by organized crime in several states, a failure of the federal government to establish an institutional framework of civic order. Nonetheless, establishing community-level political order that would also provide the base for successful community forest enterprises was a long, difficult but significantly successful process. As Fukuyama (2011, 15) notes, "As political systems develop, recognition is transferred from individuals to institutions—that is, to rules or patterns of behavior that persist over time," and this transfer happened in the agrarian sector in general and in CFEs in particular.

As in food policy, the proximate causes of reform in the forest sector were based on the reformist currents within the states, who were able to overcome opposition from local and state interests by reaching out to grassroots organizations, frequently themselves organized by the reformist elements (Fox 1992; Silva 1994). The role of intermediation in agrarian and forest policy in Mexico has also been emphasized more recently by P. Kashwan (2017). This interpretation is useful but does not take into account the unique enterprise form of the grassroots action. The reforms in the forest sector opened up not just a political space where rural people could demand more accountability from the state (Fox 2007; Kashwan 2017) but an institutional and economic space. Community forestry is not just "distributive reform" defined as "qualitative changes in the ways states allocate public resources to large social groups" or even "redistributive reform" defined as policies that "change the relative shares between groups" (Fox 1992, 10). Rather, reforms and programs in the community forestry sector, crucially including supplies of institutions and the five capitals, opened up a new political-economic and entrepreneurial space at the local territorial level. The dynamic is less between entrepreneurial reformists and autonomous social movements than between entrepreneurial reformists and an emerging communal small business sector, only occasionally and fleetingly expressed as a social movement. Once this space had been won, the sector became relatively quiet politically because it was focusing on taking care of business and like any small business sector, only became aroused when their interests were threatened (Bray and Merino-Pérez 2004). Fox notes that the food reforms of the early 1980s did not lead to lasting change. Hence, what is remarkable about the forest reforms is that they *did* create enduring structural change in the sector. The "ecological distribution conflicts" common in the periphery of the capitalist world-system create local-level resistance to "the use and abuse of nature by capital in the modern world (that) undermines not only its own conditions of production but also the conditions of livelihoods and existence of peripheral peoples" (Hornborg 2007; Martinez-Alier 2002; Orta-Martínez and Finer 2010, 216). Thus, the capitalist world-system is usually documented as having degrading effects on local communities and their environments: "local communities forfeit autonomy as they lose comprehension of events and processes that affect them" (Tainter 2007, 375). However, in hundreds of forest communities in Mexico, a different process has played out. While resistance played a historical role in varying periods, for decades many communities have been able to channel their energies toward administering their CFEs, not toward resistance. Secure forest tenure gave communities natural capital and infusions to varying degrees, and the other four capitals created a bulwark against the deleterious impacts of global capitalism, creating a space where a significant degree of political, social, and economic autonomy has been won. As introduced in chapter 1, this political-economic process has substantial implications for both common property theory and resilience theory.

EXTENSIONS OF COMMON PROPERTY THEORY

The confrontation of the empirical experience of Mexican community forestry with common property theory has revealed multiple difficulties of the theory in explaining the Mexican case. The difficulties are in the degree to which the Mexican case responds to Ostrom's design principles, the role of state policy, the role of prices and markets, the role of the source of both governance and harvest rules, and the role of the five capitals. The incorporation of the Mexican case into common property theory therefore requires an expansion of Ostrom's notion of a "facilitative political regime" and the role of markets in cases such as the Maine lobster fisheries, salmon fisheries, and Mexican CFEs. This expansion also allows common property theory to be embedded in the political economy of reform analysis carried out above.

The Mexican experience involved a long, difficult transition whereby communities of subsistence corn farmers learned to manage market-oriented industrial enterprises, in a number of cases highly sophisticated ones, through a historically and environmentally embedded process of political-economic reform. In chapter 1, I briefly reviewed a critique of Ostrom's eight design principles in light of the Mexican experience. M. Cox, Arnold, and Villamayor Tomás (2010) evaluated ninety-one studies of common-pool resources as to how well they supported the design principles empirically and theoretically. They found solid support empirically but that a number of theoretical objections had been raised. This study of Mexican CFEs finds only partial empirical support and coincides with Cox et al. (2010) that the design principles are theoretically incomplete. To this end, table 5 presents a systematic comparison of the design principles with the Mexican experience, and then, in the following table 6, I propose important principles that do explain the Mexican experience and provide important policy orientation for governments elsewhere. The rendering of the design principles are based on Ostrom's revised version, with simplified wording (Ostrom 2005).

The first design principle calls for clearly defined CPR boundaries and resource rights. Ostrom does not specify where these rights would come from, so the possibility of them coming from the state as a facilitative political regime is left open. Indeed, in the Mexican case these principles exist and come from state policy, with a nearly full bundle of rights gained over time. The second principle also has two components, related to rules over harvests and rules on distribution of benefits. As we have seen, Mexican CFEs have little to no autonomy in designing rules over harvests but have full autonomy in deciding on

TABLE 5. OSTROM'S DESIGN PRINCIPLES AND MEXICAN COMMUNITY FORESTRY

DESIGN PRINCIPLE	THE MEXICAN CASE
1. Clearly defined CPR boundaries and resource rights.	CPR boundaries established by agrarian and forest laws. CFE rightsholders have withdrawal (harvest), management and exclusion rights, but not alienation—a nearly full bundle of rights.
2. Proportional equivalence between benefits and costs. This principle includes both users devising rules on how much, when, and how products are to be allocated or harvested and rules related to distribution of benefits.	CFEs do not have autonomy to devise rules on "how much and how." These imposed rules do not crowd out incentives for collective action due to strong economic incentives. "When" is dictated by weather conditions. CFEs have full autonomy in rules related to distribution of benefits.
3. Collective choice arrangements. Most individuals affected by harvesting or protection regimes are authorized to participate in making and modifying the rules.	CFEs have little or no capacity to make or modify rules about harvests and protection of forests under management programs. Rules are almost entirely imposed by law and regulations. CFEs can make and modify rules about management of noncommercial forests, within limits.
4. Monitoring. Monitors of biophysical conditions and user behavior are at least partially accountable to the users and/or are the users.	CFE monitoring of biophysical conditions is carried out by professional foresters responding to official regulations, reflected in authorized volume of harvest. Monitoring of issues such as forest pest outbreaks is done by community. User behavior is monitored by community, though federal monitoring can apply.
5. Graduated sanctions. Users who violate rules-in-use are likely to receive graduated sanctions (depending on offense) from users, officials accountable to users, or both.	Anecdotal evidence for graduated sanctions exists, no systematic studies.
6. Conflict-resolution mechanisms. Users and officials have rapid access to low-cost, local arenas to resolve conflicts.	Community assembly and new organizational forms such as Council of Principals serve this function in CFEs.
7. Minimal recognition of rights to organize. Rights of users to devise own institutions are not challenged by external government authorities and users have long-term tenure rights.	Communities required to organize. CFE rights to organize are recognized. Long-term tenure rights are guaranteed by constitution and agrarian law.
8. Nested enterprises. Appropriation, provision, monitoring, enforcement, conflict resolution, and governance activities are organized in multiple layers.	CFEs are "nested" within clearly and legally defined multilevel governance both of the territory in general and forests in particular. CFEs may also be "nested" within regional, state, and national representative CFE organizations.

Source: Modified from Bray (2013); based on Ostrom (2005); Cox, Arnold, and S. Villamayor Tomás (2010).

allocation of benefits from the resource and in organization of the CFE (an organizational form not contemplated by Ostrom). The third design principle calls for collective choice arrangement, and again Mexican CFEs have basically no capacity to modify rules about harvests (except to harvest less than the authorized amount), though they can make and modify rules on management of noncommercial forests. The fourth design principle has the two components of social and environmental monitoring (Cox, Arnold, and Villamayor Tomás 2010), with environmental monitoring of forest productivity carried out by the professional forester and the community and some aspects of forest health, such as bark beetle outbreaks, also monitored by both. User behavior is mostly monitored by the community, though federal laws can apply. For the fifth design principle there is only anecdotal evidence, but it appears graduated sanctions are commonly used in the communities. The sixth design principle is supported by the Assembly, a state creation, buttressing earlier forms of governance in indigenous communities, and a variety of innovative organizational forms adopted by communities for CFE decision-making. There is not just minimal recognition of rights to organize but also a legal requirement for communities to organize themselves. The elaboration by Ostrom (1990, 101) of the seventh of her eight design principles is that "the rights of appropriators to devise their own institutions are not challenged by external government authorities." The perspective presented here suggests that in the contemporary world, it is necessary for government to not just not challenge but to be an actively facilitating political regime and regulate the local development of institutions and organizations (Agrawal 2001). Finally, Mexican CFEs are "nested" in a framework of multiscale governance.

Thus, Mexican CFEs actually violate some of the important design principles. As has been emphasized throughout this book, the two most important dimensions of Mexican community forests that have contributed to their success are supportive state policy and markets in a dialectic with community collective action. The state supplied institutions and organizational forms, provided models for enterprise governance that communities could then shape according to local conditions, and provided incentives for intercommunity collective action. The state supplied, to varying degrees, all of the five capitals that allowed the CFES to launch themselves. Finally, in the states where community protests and mobilizations were most important, Durango and Oaxaca, the reformist state was willing to negotiate and make concessions and use occasional direct presidential interventions in support of community forestry. The state supply of community governance institutions and organizational forms greatly reduce

TABLE 6. DESIGN PRINCIPLES OF MEXICAN CFES

DESIGN PRINCIPLE	DESIGN COMMENT
I. SUPPORTIVE STATE POLICY	
A. Devolution of rights over forest resources	Historical Sequence: 1. Long-term stable usufruct rights to community territory; 2. tree rights; 3. nearly full bundle of rights over territory and forests.
B. Supply of institutions and organizations	1. Required territorial governance institutions and organizations; 2. models for institutions and organizations for enterprise governance; 3. models and incentives for intercommunity organizations.
C. Supply of natural, social, human, physical, and financial capital	1. Natural capital supplied by rights over forest resources; 2. Social capital was variably present and other four capitals expanded or entirely supplied by state. Profitability of logging also allowed for self-financing of physical capital. Large-scale forest extension/training programs in the 1970s and 1980s were key.
D. State willingness to negotiate with and make concessions to community forest mobilizations	In Durango and Oaxaca, when forest communities organized to demand rights over forests, the state made concessions.
II. MARKET INCENTIVES	
A. Strong price incentives for collective action	Price incentives prevent crowding out of collective action around forests and CFEs despite many rules imposed from above.

Modified from Bray 2013.

transaction costs for communities of creating their own rules, though a process of social learning was required. Finally, even though most of the rules for community and forest harvest governance come from above, I argue that it is because of the price incentive that collective action is not crowded out.

What then are the design principles of Mexican community forestry? Table 6 makes a proposal. Political-economic reforms in Mexico substantially created the institutional framework that allowed for the emergence of CFEs. It devolved rights over forest resources, supplied institutions and organizational models, supplied the five capitals, and commonly acceded to demands from mobilized community forest organizations. As noted at the end of chapter 2, large-scale forest extension efforts to train community members in industrial forestry, major inputs of human and social capital, were key to the process. This infusion of support created a political-economic space where CFEs could respond to price incentives. It has been proposed that common property institutions could be replicated by simple demonstration effects. "Firms that are highly profitable are frequently used as models by others as to how to organize themselves for success. Farm households who innovate and are successful or common-property arrangements that increase their joint yield are frequently copied by others. These connections are like a ripple across the landscape rather than strongly linked connections" (Ostrom 2005, 57). But the success of many Mexican CFEs was not the subtle imagery of a ripple across the landscape or a demonstration effect. It was stimulated principally by national top-down reforms, policies and programs, and indeed "strongly linked connections" that communities were able to take and appropriate as their own. As Antinori (2000, 1) noted, entirely missing from the common property literature is a "systematic focus on stakeholders in a common property resource responding to larger market opportunities as an alternative source of benefits provided by the common property asset."

EXTENSIONS OF RESILIENCE THEORY

The century-long history of the political economy of Mexican forests has produced a sector containing hundreds of examples of forest communities and their enterprises that appear to be resilient to multiple shocks or disturbances and, crucially, to be active in mitigating and adapting to climate change. Yet, resilience theory for SES offers little explanation for this phenomenon. The successes are only very partially due to abstractly conceived multistakeholder consultations and collaboration (Goldstein 2012), the commonly mentioned sources of resilience in SES.

Human beings and their social systems in interaction with ecological systems are inherently and virtually always resilient by some measure—we respond to crises by ameliorating and adapting to the extent possible. Resilience is a fraught concept when applied to ecological systems and becomes even more so when applied to SES. One the one hand, the heuristic power of the concept makes it indispensable in discussions of sustainability (Anderies et al. 2013). On the other hand, the empirical resilience, based on decades of survival, of the Mexican community forestry SES is based on aspects not discussed in resilience theory. There are three reasons for community forestry's political resilience: (1) it has had consistent public policy support, of varying intensity, since the early 1970s and even earlier in some cases, (2) it has a strong template and foundation of community and enterprise governance in Mexican law, and (3) it has the capacity to engage in collective action to protect its interests and expand its economic and political space. It is economically resilient because communities own a territory with a valuable natural resource that gets consistently high prices in the market. It is socially resilient because the low transaction costs of the existing governance institutions and high market value facilitate collective action that strengthens existing social and capital and builds social capital where it hardly existed. It is ecologically resilient because both government regulation and community norms and intergenerational values converge on managing for sustainable forest stewardship, now augmented by an emergent community culture of conservation. As will be expanded further below, it is resilient to climate change both because good forest management mitigates climate change by maintaining forests as carbon reservoirs and by harvesting carbon stored in timber for long periods in the form of construction materials and furniture. Specific adaptations to climate change are only starting to emerge, but will become increasingly important to implement in the coming years. Resilience has emerged due to a complex historical process involving strong reform actors from above in interaction with strong community response and mobilizations; the exact sequencing of these top-down and bottom-up forcing variables have varied from region to region in Mexico. Resilience is currently most challenged by the failure of the Mexican state to control organized crime and to maintain consistent policy support in recent years.

Almost none of these empirically observed dimensions of resilience in the Mexican forest SES are captured in most of resilience theory. Theories of forest resilience in the face of climate change focus mostly on actions that managers should perform, framed as purely technical decisions (Messier et al. 2015; Millar, Stephenson, and Stephens 2007). For example, forests should be managed for "species persistence within large ecoregions" and for "creating porous landscapes" (Millar, Stephenson, and Stephens 2007, 2150). Forests should be viewed as complex adaptive systems that will putatively help "increase the adaptive capacity of forest ecosystems" by planning and accessing "interventions across a range of spatial and temporal scales, e.g., from plant neighborhoods to landscapes" (Messier et al. 2015, 373–74). These technocratic calls seldom deal with the messy social and political landscapes in which most forests actually exist and focus only on the forest, not a forest SES. Efforts to advance the concept of a resilient community forest SES are also limited. W. N. Adger (2000, 361) defines social resilience as "the ability of communities to withstand external shocks to their social infrastructure" in the face of environmental, political, and economic shocks, enhanced by social learning. Berkes and Ross (2013, 17) suggest that "social capital and networks, sense of place, values, and social identity are key to community resilience." The resilience of the Mexican forest SES is built upon much more than these important but limited factors.

Discussions of what constitutes a resilient social-ecological system do not do much more to help us understand the Mexican forestry SES. The concepts of C. Folke (2006) of "adaptability" and "transformability" were briefly discussed in chapter 1. Adaptability is the "the capacity of people to build resilience in social-ecological system through collective action" in spite of disturbances. Transformability is "the capacity of people to create a fundamentally new social-ecological system when ecological, political, social or economic conditions make the existing system untenable" (Folke 2006, 263; Walker et al. 2004). The former suggests a dynamic state that is stable within a given "basin of attraction," whereas the latter suggests a transition to a very different alternative stable state that may be more optimal.

Mexican community forestry can be conceptualized as having passed through a transformation from a stable state of government-directed forestry to a stable state, a different basin of attraction, of community-directed community forestry, which is indeed a "fundamentally new social-ecological system." Its survival certainly shows adaptability to ongoing political and economic disturbances, for instance, shifts in government policy, bilateral agreements such as NAFTA, and marketing issues and ecological disturbances from bark beetle infestations in the temperate forests and the looming challenge of climate change. However, some of the fundamental problems with resilience theory are similar to those with institutional theory. The transformative decentralization of Mexican community forestry has been a decades-long process driven by much more than social learning and social capital, though both are certainly present. As well, where does the "capacity" that Folke attributes to an abstract "people" come from? "Capacity" does not just magically emerge in response to a disturbance. The ecological metaphor of disturbance is seriously limited, because what enabled "people" to act collectively in the case of Mexican forests were state policies, market forces, the supply of the five capitals, and indeed community social capital, all of which may alternately constrain or encourage transformations and adaptation.

MEXICAN COMMUNITY FORESTRY IN AN INTERNATIONAL COMPARATIVE FRAMEWORK

In addition to challenging common property and resilience theory, the Mexican case also challenges the empirical generalizations of a number of large-N studies on the variables that lead to success on the commons. The most quantitatively based empirical work on the dynamics of forest commons emerged from Ostrom's signature project, the International Forestry Resources and Institutions Program (IFRI). The IFRI project has a database on social and ecological variables in forest management on some 200 communities in ten countries (Agrawal and Chhatre 2006; Chhatre and Agrawal 2008; Gibson, Williams, and E. Ostrom 2005; Persha, Agrawal, and Chhatre 2011). Studies drawing on the database, collected with a standardized methodology, have examined clusters of biophysical, demographic, economic, institutional, and sociopolitical variables that influence the broad ecological variable of "forest conditions." Forest conditions are measured by stakeholder perceptions of degradation or regeneration, remote sensing, measurements of tree stem diameter and density, and tree species richness (Randolph et al. 2005). These studies have found that governance dimensions, including varying combinations of local monitoring and enforcement of rules, collective action around forests (e.g. in reforestation), commercial versus subsistence dependence on forests (based on harvesting of firewood and fodder), size of forests, and forest altitude/type can all influence forest conditions. Highly precise findings include, in an analysis of 152 forests across nine countries, that "forests with a higher probability of regeneration are likely to be small to medium in size with low levels of subsistence dependence, low commercial value, high levels of local enforcement, and strong collective action for improving the quality of the forest" and that "the probability of degradation increases with the commercial value of forests" (Chhatre and Agrawal 2008, 13287). Furthermore, "For forests . . . very large territories are unlikely to be self-organized given the high costs of defining boundaries (e.g., surrounding with markers or fences), monitoring use patterns, and gaining ecological knowledge. Very small territories do not generate substantial flows of valuable products. Thus, moderate territorial size is most conducive to self-organization"

(Ostrom 2009, 15). Persha, Agrawal, and Chhatre (2011) found in an analysis of eighty-four community forests in six countries that local rule-making is the most important variable in determining jointly positive outcomes of forest subsistence livelihoods and biodiversity conservation but that forest size and extractive commercial livelihoods are also important.

Research on Mexican community forestry only minimally bear out the findings from these large N studies. Of the successful examples of Mexican community forests in all size ranges, from a few hundred hectares to over 100,000 hectares, all are managed for their commercial, not subsistence value, and local rule-making over the resource has almost no role in success on the Mexican forest commons. High commercial value of forests in Mexico is due to strong market connections and has led to forests that are not degraded but are constantly regenerated through state-regulated silvicultural practices. However, it is certainly true that high levels of local enforcement and strong collective action are commonly present, but around both industrial forestry and improving forest quality.

With some 60 percent of its forests owned by communities, Mexico is a global outlier here as well. As of 2017, an analysis of trends in forty-one countries shows that just over 15 percent of total forest area in those countries was legally owned by or designated for indigenous peoples and local communities, an increase of only 5.6 percent since 2013. Although discouragingly small, the increase in community tenure between 2013 and 2017 is due to increases in community forest direct ownership, with over 90 percent of this progress stemming from low- and middle-income countries in Africa, Asia, and Latin America. Recent laws in a number of countries have established new legal pathways for communities to own their forests under national law, but they currently fall far short of Mexican accomplishments (Rights and Resources Initiative 2018).

These advances, however, refer only to tenure rights, and don't say anything about how the forested lands are managed. An early global review by D. J. Klooster and S. Ambinakudige (2005, 324) concluded that "Mexico is in the vanguard of the community forestry movement worldwide on several fronts." The authors note that community ownership of forests is more extensive than anywhere else in the world, tenure is secure, and rights to harvest and sell trees exist. Furthermore, despite a strong regulatory framework Mexican CFEs have central roles in forest management, and ownership is based on local governance institutions that are generally democratic and autonomous. After comparing forestland regimes in India, Tanzania, and Mexico, P. Kashwan (2017) concluded that forest and conservation policies in India and Tanzania largely exclude forest-dependent peoples, while they are central to these policies in Mexico.

Beyond Mexico, possibly the most mature example based on the production of timber is in the Petén of Guatemala. The experience in the Petén is based on twenty-five-year concession contracts between the state and organized groups within communities (Bray et al. 2008; Monterroso and Barry 2012; Taylor 2010). The community concessions give significant use, access, management, and exclusion rights to all forest products for twenty-five years with the requirement that they receive FSC certification for sustainable management. The situation of many of the community concessions in the Petén was much more challenging than in many Mexican forest communities. The twelve community concessions (there were also two industrial concessions) had varied circumstances, including nonresident community concessions, concessions to communities of recent colonists, and concessions to long-inhabited areas historically focused on NTFP production. They have struggled to varying degrees with poor financial management, high turnover in management positions, and internal and external conflicts, and some have become inactive (Radachowsky et al. 2012). However, a recent report finds that the nine active concessions have shown a process of growth, professionalization, and consolidation; have extremely low deforestation in a region with high deforestation; brought certified forest management to 352,089 hectares; and had a gross income of $US24.7 million from 2012 to 2016, among other accomplishments (Stoian et al. 2018). Although more reduced in scale than in Mexico, this production record is a strong demonstration that devolving significant rights over forest resources can be successful in a context very different from Mexico.

The other large-scale, but still little studied, experience in community forest management for timber has taken place in Bolivia. Legal reforms in 1996 created the new figure of Community Origen Lands (Tierras Comunitarias de Origen; TCO) and gave the TCO communities the rights to manage their forests for timber and NTFPs. By 2006, eighty-eight indigenous CFEs had been initiated, forty-eight of them with approved management plans covering 1.345 million hectares. Serious governance problems over enormous territories, without the kinds of multilevel governance institutions established by the Mexican experience, have impeded the development of the CFEs, though there are a few examples of relative success (Benneker 2008; Cronkleton, Bray, and Medina 2011). The impacts on community incomes and welfare and on forest cover of the Bolivian CFEs have been understudied and merit greater attention.

In most of the rest of the world, community forest management is focused on NTFPs, giving communities limited rights to resources of limited value with, not surprisingly, limited results. Joint Forest Management in India and community forestry in Nepal mostly give rights over modest acreage of degraded forest lands that are improved by the community to produce crucial subsistence needs such as fuelwood and fodder but that do not permit economic development. The 2006 Forest Rights Act (FRA) in India in principle empowered rights holders and village institutions to use, manage, govern, and conserve forests and biodiversity in forests traditionally used by the communities (Kashwan 2017). But due to entrenched bureaucratic power, the most advanced state, Maharashtra, has only 20 percent of its forests recognized under the FRA (CFR-LA 2016; Kashwan 2017). As of late 2018 both individual and community forest rights were recognized nationally in only 15.29 percent of the potential forest area and the average recognized area for Community Forest Rights across India was only 127.78 acres (Sahoo and Sahu 2019), a scale far smaller than in Mexico. In most of these cases, rights have been given to low-value NTFPs such as bamboo. In Nepal, the National Forestry Plan of 1976 started a policy shift toward participatory management that was consolidated by the Forest Act of 1993. Community Forest User Groups (CFUG) are self-governing corporate bodies with legal recognition and are authorized to make management decisions over forests on the basis of a government-approved management plan (Acharaya 2002; Gautam, Shavakoti, and Webb 2004). As of 2009, 14,439 CFUGs managed 1.23 million hectares, around 22 percent of national forests. However, these outcomes starkly contrast with Mexico, since the average size of a community forest in Nepal is 82.7 hectares and most production has been for subsistence purposes (Raj Sharma 2009).

The most important example of community forestry in Africa is in Tanzania. There, community resistance to the declaration of protected areas in the early 1990s eventually led to the National Forest Policy and Forest Act of 2002, which provided the legal basis for Participatory Forest Management (PFM) that allows communities to own, manage, or comanage forests under a wide range of conditions and that recognizes village governance structures as formal corporate entities that can own property (Blomley and Iddi 2009; Kashwan 2017). There are two major forms of agreements under PFM: lease rights over government reserves through Joint Forest Management (JFM) agreements or ownership of forests on government or village lands through Community-based Forest Management (CBFM) agreements. An estimated 2,300 communities manage some four million hectares, or around 11 percent of the forests. In Tanzania, the majority of CBFM agreements are in the dry miombo and acacia forests, while JFM agreements are concentrated in the evergreen forests of the Eastern Arc Mountains. There are also some seventeen million acres of "unreserved forests" in poor remote areas with the potential for commercial timber production (Blomley and Iddi 2009). In Tanzania the 1,460 communities with CBFM management rights have 2.35 million hectares for an average of 1,609 hectares (calculated from Blomley and Iddi 2009, table 2), but nearly all are used for subsistence purposes or low-value NTFPs.

The fact these other experiences globally generally involve much smaller forests and lower value NTFPs does not mean that they don't constitute significant accomplishments in the contexts in which they occur. Macqueen et al. (2018) have shown how many examples of "locally controlled forestry" in Nepal and Tanzania and elsewhere represent organizational innovations and meet a broad definition of increased prosperity. Nonetheless, the degree of state support, the access to high-value timber, and the degree of vigorous community response and initiatives highlight the degree to which Mexico is a global standout of "success on the commons" (Chhatre and Agrawal 2008) with important lessons for governments elsewhere.

PROSPECTS FOR THE FUTURE

As of early 2019, the new government of President Andrés Manuel López Obrador had brought community forest management to the center of the forest policy stage, positioning it as a central strategy in combatting deforestation, though subsequent budget cuts substantially reduced CONAFOR subsidies to the sector. In any event, can the sector, whether as an instrument for braking deforestation or combatting climate change, really be successfully expanded to include significant numbers of new forest communities? The prospects are not clear. The policy initiatives of the last half of the Calderón presidency (2013–16) were substantially oriented toward promoting new CFEs and the upward mobility of existing ones. The results suggest substantial barriers in both launching new CFEs or preventing downward mobility in vertical integration. Torres Rojo and Graf Montero (2015) have defined a long list of variables that can influence the level of vertical integration, including the quantity and quality of the resource, the maturity of governance of the community, technical and management capacity, market access, and the particular social and economic environment of the community, among others. The mortality, or going out of business entirely, of CFEs at any level of vertical integration can be variously associated with poverty, lack of investment, community conflicts, or a rapid liquidation of timber reserves (Torres Rojo and Graf Montero 2015). This list suggests substantial impediments to progress that would require much larger investments to overcome. Policy initiatives to increase levels of vertical integration met with modest success, but "It is now understood that there is an optimal level of vertical integration and that in spite of efforts to achieve higher levels of integration, it is difficult to break the restrictions placed by the indicated variables. Thus, each CFE should achieve the level of integration that agrees with its physical, economic, social, cultural and contextual potential" (230).

The related question is, "How feasible is it to bring new communities, Type I communities who have never logged, into the forest production sector? Has the available pool of likely candidates to become CFEs become mostly exhausted after decades of efforts, with the sector at its natural limits?" The many thousands of communities with temperate and tropical forests shown by statistics do not indicate anything about the size or commercial quality of those forests. The temperate forests could be dominated by noncommercial oaks or degraded to various degrees. In tropical forests, given the lower density of commercial stems, relatively large forests are required to be commercially viable. For communities who may have commercially viable forests but who are not logging them, there has been a lack of information on why they do not log or how many there are. CONAFOR (2012) suggest that agrarian problems, organizational issues, problems with access to forest areas, technology and markets for processing and marketing hardwoods, technical assistance, and others issues have impeded the incorporation of these areas to forest management. J. M. Torres Rojo and S. Graf Montero (2015) note that in trying to organize new CFEs, CONAFOR made the assumption that starting with human capital formation in silviculture, forest management, administration, and related issues would be the best strategy. However, the commission found that this "was only valid in those agrarian communities that already had a good level of organization and had developed a governance structure that permitted efficient decision making" (231).

Hernández-Aguilar et al. (2017) shed some useful light on the issue of the limits to establishing new CFEs. The Mixteca region of Oaxaca is not known for the presence of CFEs. In a study of twenty-five forest communities in that region with more than 300 hectares of forest, most of them with much more, the researchers found that only three of the twenty-five were operating CFEs for timber production. Another six were producing NTFPs (pine resin or bottled water) and had no interest in timber production. The other sixteen had no legal forest production at all, and there appeared to be both biophysical and social factors at work. The nonproduction communities tended to have lower precipitation and poorer soils than the CFE communities, so likely less productive forests. As well, they had weak governance, poor enforcement of existing land-use rules, and high rates of emigration. It is likely that large numbers of communities with commercial-sized forests but no production exhibit similar characteristics. Given the concerted effort of the government programs to launch new CFEs in the last three years of the Calderón presidency, it is possible that Mexican forest policy has been bumping up against the practical limits of further expansion of community forest management without major investments in extension efforts. It is frequently argued that overregulation is the most important barrier to the expansion of the CFE sector (CCMSS 2006; G. Chapela y Mendoza 2018), and while this is clearly a factor, there are also evidently many other factors that impede more dynamic development. Some of the limits discussed above could conceivably be overcome in a number of cases, but it will require very large new investments in four of the five capitals (assuming existing natural capital) and marketing support. A well-designed program would likely be able to continue to find at least some targets of opportunity to create new CFEs and help existing ones climb the value-added ladder of vertical integration, though major further expansion of the sector does not appear likely (interview with S. Anta, March 20, 2017). In the meantime, it appears that in many hundreds of cases, Mexican CFEs will be able to continue on their path of generating incomes and maintaining forest cover and biodiversity, and staying resilient on multiple fronts but most importantly against climate change.

CLIMATE CHANGE AND REDD+: MEXICAN FORESTS ON THE INTERNATIONAL STAGE

While it is not clear how much the model can be expanded in Mexico without major new investments, the large existing sector is a dynamically adapting SES that is now beginning to face the major and persisting environmental disturbance of climate change. In the 1970s, when the community forest sector began organizing as an SES, anthropogenic climate change was only beginning to be studied by scientists, but it is now an increasingly evident and present reality, and, as noted in chapter 1, in the future we will need to talk about the SECS of Mexican community forestry. The community forestry SES or SECS is beginning to experience the new stressor of climate change from higher precipitation, higher temperatures, and extreme storm events, as well as increasing attacks by bark beetles (*Dendroctonus spp.*) and other pests driving threats to rural livelihoods and loss of biodiversity (Eakin, Lemos, and Nelson 2012; Estes et al. 2011; Durán and Poloni 2014; Rands et al. 2010; Tucker et al. 2010). Climate change models for the forests of Mexico suggest an expansion of arid biomes such as tropical dry deciduous forests and Sonoran Desert scrub that will progressively replace conifer forests, semideciduous forests, and cloud forests (Rehfeldt et al. 2012), the conifer and semideciduous forests being where most CFEs operate, as we have seen.

These emerging climate stressors have driven widespread concern over the possibility of catastrophic shifts in forest ecosystems (Donangelo et al. 2010; Rietkerk et al. 2004; Scheffer et al. 2001; Scheffer et al. 2012). In the Amazon basin, interactions between deforestation, fire, and climate-driven drought are feared to be leading to a "tipping point" of rapid degradation of the Amazon forest (Nepstad et al. 2008). M. Rietkerk et al. (2004) have suggested that an indicator of an impending catastrophic shift in ecosystems may be "self-organized patchiness" of consumers and resources in an ecosystem. There is solid evidence that Mexican CFEs are making substantial contributions to resilience against catastrophic "tipping points" and mitigation of and adaptation to climate change in what could be called "self-organized intactness." Many community forest regions of Mexico are characterized by an apparent stability and linearity or "adaptability" of the processes of supportive public policy, price stability for timber, community governance, forest-based incomes, and intact forest cover and forest ecosystems. That is, they are stable within their basin of attraction. They are guided by stabilizing positive feedback loops from strong community governance, supportive public policy, and market incentives for collective action around timber management, as suggested in the conceptual framework introduced in chapter 1. They show a marked tendency toward the organization of extensive income-generating and intact forest masses.

Thus, Mexican forests provide an important model for global forests, and one that emerged onto the world stage in the context of REDD+. REDD+ refers to the Reduction of Deforestation and Degradation, plus conservation, sustainable management of forests, and enhancement of carbon stocks. As noted in chapter 1, it is an international framework that incorporated forests into efforts to reduce carbon emissions, was negotiated under the UN Framework Convention on Climate Change (UNFCCC) beginning in 2005, and was endorsed as part of the Paris Accords in 2015. The concept is that industrialized countries will provide financial incentives through forest carbon markets to developing countries for reducing emissions from forest loss and degradation as "results-based finance" (Seymour and Busch 2016). Forests were not included in climate negotiations for many years for a variety of political reasons and concerns about "permanence" (carbon stored in forests can easily go up in smoke) and "leakage" (forests conserved in one place could result in forests being cut down elsewhere). Forests finally began to emerge from the international political wilderness at the Conference of Parties (COP) negotiations in Montreal in 2005 with a proposal from Papua New Guinea and Costa Rica then called Reducing Emissions from Deforestation (RED). At the 2007 COP in Bali, thanks to further efforts from these two countries, forests were formally included in the Bali Action Plan. The Bali Action Plan proposed financial compensation for avoided forest carbon emissions and called for the development of technological capacity to monitor deforestation and demonstration projects (Seymour and Busch 2016). It was hoped that the COP meeting in Copenhagen in 2009 would achieve a comprehensive climate agreement and ambitious emissions reduction targets, including large-scale finance for forests as part of the agreement. The collapse of the negotiations and a fig leaf of an agreement did, however, include REDD+ as a component of future negotiations and called for the establishment of national forest monitoring systems.

It was in this context of this international activity that Mexican forests officially entered the realm of international politics on climate change in 2010 at the COP 16 meetings held in Cancún. In late 2010, the Mexican government laid the groundwork for the national REDD+ program with the document *Visión de México sobre REDD+ y la Estrategia Nacional REDD+* (Mexico's Vision on REDD+ and the National REDD+ Strategy; CONAFOR 2010). In this document, Mexico declared that "sustainable rural development" was to be the central animating REDD+ strategy. Mexican community forests were highlighted at the COP 16, particularly through the PHS program, through a major speech by then-president Felipe Calderón. The carbon generated by the travel to the conference was offset by purchasing forest carbon credits from the Oaxaca-based NGO Environmental Services of Oaxaca (Servicios Ambientales de Oaxaca; SAO) (Bray 2012). With funds from the World Bank's Forest and Climate Change and Forest Investment Programs, Mexico established the goal of administering its forests sustainably, creating social capital around their protection, practicing sustainable use, and generating additional income from forest products and services, including REDD+ (CONAFOR 2010).

The Mexican REDD+ program began with the so-called Special Programs for REDD+ Early Action Areas in 2010–14 and focused on the coast of Jalisco, the Peninsula of Yucatán, and the Lacandón rainforest. These programs were complemented by a U.S. Agency for International Development / Mexico Global Climate Change Program (2011–15) called Mexico Reducing Emissions from Deforestation and Forest Degradation Program (MREDD) that sought to strengthen policies, laws, and institutional and technical capacity; create the necessary financial architecture; and establish internationally robust monitoring, reporting, and verification (MRV) systems for REDD+. A review of the Special Programs was highly critical and found that it had temporarily reduced deforestation but that further interventions were needed to assure the permanence of forest cover in the long term (Deschamps, Zavariz Romero, and Zúñiga Pérez-Tejada 2015). More than half of all funds were spent in the state of Jalisco, and the PES program was the primary instrument, with only 10 percent of funding going to community forestry. Some regions were chosen because they were strongholds of the political party in power (Trench et al. 2018). The program was considered to be a unique opportunity to improve forest policy, but multiple barriers impeded the realization of its goals. Lack of interest and of political will perpetuated a more business-as-usual approach by CONAFOR, institutional follow-up was scarce, and mechanisms to incorporate learning were missing. There was also little coordination between the relevant federal agencies, little attention to governance issues, and no mechanisms for monitoring deforestation (Trench et al. 2018).

However, the good news was that despite these uncertain accomplishments, and efforts that substantially stalled out during the Peña Nieto government, community forestry continued to stand as a model of sustainable rural development in Mexico. Indeed, the current outcomes in many, but not all, regions of Mexico can be thought of as "post-REDD+ landscapes"—that is, what landscapes elsewhere in the world will look like if REDD+ strategies to reduce deforestation and achieve cobenefits of sustainable livelihoods and biodiversity are successful (Bray 2010a). Since studies show that many forests globally have an increased risk of drought and heat-induced tree mortality (Allen et al. 2010), the managed forests of Mexico are a crucial model for a strategy that may reduce these risks. The post-REDD+ dimensions of Mexican forests point the way for forest and human community resilience that can both mitigate and adapt to climate change.

MEXICAN CFES AND MITIGATION OF CLIMATE CHANGE

A number of studies have found that Mexican forests can make substantial contributions to carbon capture. As early as the mid-1990s it was conservatively estimated that the total economic value of Mexican forests was around US$4 billion, nearly 95 percent of the value coming from carbon storage (Adger et al. 1995). O. R. Masera, J. Ordóñez, and R. Dirzo (1997) projected a policy scenario in which the forests of Mexico could become a carbon sink by 2025, and this has actually happened in the temperate forests, as we saw in chapter 7. D. Klooster and O. Masera (2000) argued for the relevance of Mexican community forestry for carbon mitigation within the Clean Development Mechanism in the 1990s. More recently, a 2018 study of carbon storage in above- and belowground dry woody mass and soil organic matter in sixty-four countries in all four major biomes (tropical, subtropical, temperate, and boreal) found that Mexico had the fifth-largest carbon storage globally (Frechette, Ginsburg, and Walker 2018). Mexico also has some of the best-documented projects in the voluntary carbon market globally. These include an agreement with the Oaxaca NGO Integrator of Indigenous and Peasant Communities of Oaxaca (Integradora de Comunidades Indígenas y Campesinas de Oaxaca; ICICO) and the community of San Juan Lachao to offset the carbon of major Mexican corporations, the city of Palo Alto, California, through California's Climate Action Reserve (CAR), as well as the Walt Disney Company and Duke University (Osborne and Shapiro-Garza 2018; E. Shapiro-Garza, personal communication, September 9, 2019). Beyond Mexico, B. W. Griscom et al. (2017) calculated that what they call "natural climate solutions" can provide 37 percent of cost-effective carbon dioxide mitigation needed through 2030 to have a >66 percent chance of holding global warming to below 2 degrees Celsius. They consider twenty conservation, restoration, and improved land management options that increase carbon storage and/or avoid greenhouse gas emissions, constrained by the need for food and fiber security and biodiversity conservation. Of the twenty, the forest options have over two-thirds of potential mitigation, and the third most important of these is "Natural Forest Management," which offers "large and cost-effective mitigation opportunities" (Griscom et al. 2017, 4). This argues for large and cost-effective opportunities for mitigation in Mexico's community forests.

As we saw in chapter 7, the majority of the forests in the forest management communities are in varying degrees of conservation, where carbon capture is an ongoing and protected process. However, the fact that significant parts of the forest are under management for timber actually gives them an enhanced value for carbon storage, since most of the harvested timber goes into the long-lived forest products pool or "slow pool." We reviewed silvicultural practices in Mexican forests in chapter 7, but here I will review their implications for carbon (C) capture and mitigation of climate change (Bray and Duran-Medina 2014). One pertinent question is if there are quantifiable differences in long-term C capture between the uneven-aged and even-aged silvicultural practices (MMOBI and MDS and its variants) dominant in Mexican silviculture. Long-term C capture is dependent upon both the rates of biomass accumulation under each practice and on harvest impacts that may release carbon (as discussed in chapter 7).

Logging of forests creates a cyclical pattern of carbon release and capture whereby intensively managed stands store less carbon than unmanaged forests, though the degree of carbon storage also depends on the extent of the forest under study (Tyrell, Ross, and Kelty 2012). For example, a study of carbon fluxes in the managed forests of Ixtlán de Juárez, Oaxaca, found that 63 percent of the baseline carbon stock was removed in around 1,200 hectares (largely due to the intensive silvicultural practices mentioned in chapter 7), but eight years later, 11.7 percent of the extracted carbon had been recovered in an ongoing process of carbon capture, and Ixtlán has over 10,000 hectares of forest under conservation (Pacheco-Aquino, Durán Medina, and Ordóñez-Díaz 2015).

The amount of carbon released and captured and the time period over which it occurs can vary greatly between and within different silvicultural treatments, and the final destination of harvested trees is important. D. C. Bragg and J. M. Guldin (2010) have defined as "fast pool" biomass smaller roots, bark, foliage, and other slash that decomposes quickly and tree products that are converted into short-lived paper products. "Slow pool" biomass consists of stumps or larger pieces of the trunk that may be left behind and timber in long-term storage in furniture and buildings (Carroll et al. 2012). In the Mexican case, most of the harvest goes into the slow pool in long-lived forest products, and the requirement to chop slash and arrange it in contours to prevent erosion likely significantly reduces carbon emissions from soil disturbances and the decomposition of the slash. Uneven-aged and even-aged silvicultural systems have tradeoffs with respect to emissions, and which is superior depends largely on the specific management practices and the time horizon (Bragg and Guldin 2010). A study of forest management and carbon sequestration in Oaxaca found that clear-cutting in forty- to fifty-year cycles would be superior for C capture to the group-selection practices now used by some Sierra Juárez communities (Alvarez and Rubio 2016), but regulatory norms do not currently permit this practice.

The delayed harvest of what is the joint production of timber and carbon (Amacher, Ollikainen, and Koskela 2009) is common among Mexican CFEs, for a variety of reasons. A study of CFEs in Oaxaca and Chihuahua documented that in both states it is common that the complete authorized volume for a logging year is not harvested, due to late-arriving permits, to rain, to disorganization, or to conservation-oriented decisions to not harvest all of the volume (as noted in chapter 7). When this happens, the unlogged volume does not carry over to the next logging year, by government regulations. What takes place, then, is a de facto partial lengthening of the harvest cycle, since that volume would not be harvested for another ten years (Bray and Duran-Medina 2014). As a result, this practice is functionally carbon enhancement.

MEXICAN CFES AND ADAPTATION TO CLIMATE CHANGE

Bark beetles and some other pests are currently the only impacts likely related to climate change evident in Mexico's forests as of the late 2010s, but further serious impacts are projected. L. Gómez-Mendoza and L. Arriaga (2007) use climate change models to show that the current geographic distribution of pines could be reduced anywhere from 0.2 to 64 percent depending on the species, including major reductions of some important commercial species. Sáenz-Romero et al. (2010) also use modeling to predict increases in temperatures and reductions in precipitation that will drive a general migration of vegetation toward higher elevations and will likely affect the growth of pines by 2030. Both studies suggest that pine species at lower altitudes, such as *Pinus oocarpa*, and pines at higher elevations, such as *Pinus hartwegii*, are particularly vulnerable. The altitudinal requirements for *Pinus oocarpa*, important for resin, could be as many as 300 meters higher in 2030 than currently. For Michoacán, the authors propose a program of assisted migration for important commercial pine species. A similar proposal for assisted migration of the important commercial pine species *Pinus patula* has been made for the Oaxaca community of Ixtlán de Juárez. However, in late 2018 I had a conversation with the forester for the community, and he reported that *Pinus patula* was observed to be growing vigorously at both low and high altitudes, so no impact of climate change on that species had yet been observed (Ruiz-Talonia et al. 2014; interview with E. Santiago García, September 12, 2018). A recent report by the Intergovernmental Panel on Climate Change (2018) on strategies for keeping global warming below 1.5 degrees Celsius—that is, below preindustrial levels—proposed the following: avoidance of forest degradation and deforestation, usage of local and indigenous knowledge, application of biodiversity management, and usage of community-based adaptation, among other options. However, these findings neglect to include a key adaptation option of Mexican community forestry, in addition to local and indigenous knowledge: the encouragement of new community cultures of industrial forestry applied to forest management.

Indeed, the model of Mexican CFEs can be considered as a preadaptation to impacts of climate change. I have detailed the multiple ways in which Mexican CFEs can be considered to be resilient. Similarly, the CFEs' political, economic, social, and ecological resilience makes them resilient to climate change. Support to Mexican CFEs seems firmly entrenched in Mexican forest policy for the foreseeable future. Recessions may impact the sale of timber, but timber is a product that will always have some demand, so the incentives for collective action are unlikely to be greatly reduced. Many forest communities have such high forest cover that they could easily increase food production if necessary without harming commercial timber harvest. Thus, community collective action around common property forests in Mexico is posited to be a frontline adaptation to modifying the rate and direction of forest ecosystem response to climate change (Burkett et al. 2005; Diffenbaugh and Field 2013). The Mexican model provides inspiration for how many forest-rich developing countries can increase the resilience to climate change of their forest communities.

THE SEEDS OF A GOOD ANTHROPOCENE

As discussed earlier, multiple characteristics of Mexican common property forest institutions force a substantial rethinking of theory and policy. Mexican community forestry represents an outlier case of a "virtuous circle" of outcomes on the forest commons (George and Bennett 2005) that require an extension of existing theory. State policy as a facilitative regime and markets have actually strengthened collective action around the forest commons, and these factors have driven multiple forms of capital inputs required by enterprises, creating new forms of state-supported "self-organization" in community forest enterprises and intercommunity forest associations (Antinori and Bray 2005; Bray, Antinori, and Torres-Rojo 2006; Bray, Durán, and O. Molina-González 2012). These factors created a "pathway to transformative change" that have positioned Mexican community forestry to make an important contribution to a "good Anthropocene" (Bennett et al. 2016, 441).

Mexican CFEs run the gamut from those that have been astoundingly successful to those that are besieged by factors that range from small forests with low economic value and community disorganization to assaults by organized crime. For the successful ones, San Pedro el Alto in Oaxaca, profiled in chapter 4, may well be at one extreme, expressing what Garibay Orozco (2008, 438–39) has termed "a kind of realization on the earth of the purest values of peasant communalism. Especially the communalism idealized by ethnic social movements: the community understood as a harmonic social space, autonomous, self-governed, territorial, egalitarian, collectivist, prosperous, rooted in the values of tradition and customs." As noted, this degree of egalitarian democratic decision-making makes the village democracies of New England pale in comparison. At another extreme is the case of El Balcón, where a CFE that was highly successful for some three decades is now besieged by tensions and violence emanating from failures of the state to establish law and order in rural Mexico. The recent example of Cherán in Michoacán, where the community had to organize in the midst of violence to expel illegal loggers, is also representative of the struggles of many Mexican communities (Boyer 2015; Zabludovsky 2012), but with a more positive recent outcome.

The multiple inefficiencies of CFEs as enterprises and a burdensome regulatory apparatus will continue to present challenges, but mostly ones that have now survived for decades. But beyond these issues, the biggest single threat to Mexican CFEs in a number of states is organized crime and the failure of the state to control it. CFEs in states such as Michoacán, Durango, Chihuahua, and others have experienced impacts ranging from having to pay protection money to having the CFE directly taken over by organized crime.

Despite these major challenges, Mexican CFEs are now an experiment that has run over fifty years with proven results in hundreds of cases. Mexican communities have shown creativity in designing rules and organizational innovations around the CFEs, alongside significant historic support from the state in the top-down institutional supply of rights and rules and the five capitals. More data is needed on economic impacts, but it is clear that where CFEs are sufficiently large, there are significant employment, benefits, and likely poverty alleviation outcomes. A major cultural transformation has taken place in hundreds of communities: a culture of industrial forestry has displaced and/or augmented many aspects of traditional cultures of production and forest use. The constant rotation through leadership positions established by agrarian law and community norms in indigenous areas has led to most legal members of the community having a basic working knowledge of many aspects of industrial forestry, from silviculture to accounting.

The proposal of this book is that Mexican community forests and their CFEs present the best case in the developing world for how a forest-based SES has evolved toward resiliency, mitigation, and adaptation in the face of climate change for interdependent forests and communities. J. B. Alcorn and V. M. Toledo (1998) correctly point out the foundational importance of property rights to resilient community forest management in Mexico. However, it is a range of other large-scale state policy initiatives and market incentives that have been crucial to the evolution of the Mexican community forestry SES. State policy, both unintentionally and intentionally, opened up the spaces for political mobilizations and community entrepreneurial vigor to occupy and expand those spaces, and in providing all or significant measures of the natural, physical, financial, human, and social capital necessary to create market-viable CFEs. The market incentive of consistently high prices has ensured that there is a permanent motivation to incur the substantial transaction costs of engaging in the collective action to organize a CFE. As depicted in figure 1 in chapter 1, in the 1970s and 1980s institutional regime shifts toward collective action around CFEs have generated positive feedback mechanisms from community collective action, government policy, and markets, towards expansion of forest cover, biodiversity conservation, and sustainable livelihoods, that is, a more desirable alternative stable state of forest governance (Kinzig et al. 2006). CFEs appear to be well buffered against catastrophic shifts in a resilient coupled social-ecological system. They also demonstrate that community forests can be as effective as public protected areas in conserving forest cover and biodiversity, while also generating income for local communities.

In an era of widespread concerns about catastrophic tipping points in ecosystems at multiple scales, community forest regions of Mexico provide a model for successful land-use governance in tropical regions and as "safe and just" operating spaces for a regional and national SES (Dearing et al. 2014). The evolution of the SES has been driven by policy designs and a "knowledge infrastructure" that have engaged a diverse set of "actors, aspirations, and beliefs" toward the safe operating space or defensive perimeter for the emergence of the CFEs (Anderies, Mathias, and Janssen 2019). The degree to which this accomplishment is true depends upon many factors discussed in this book, and clearly not all forest regions and CFEs in Mexico can be characterized as successful, safe, and just, but the significant number that are shows the enormous potential of the model. To further consolidate and expand the sector and in addition to the significant and important physical capital subsidies provided by CONAFOR, a larger program of sustained forest extensionism is required to build human and social capital where it is absent. In a 2018 policy statement by Mexican forest community organizations and academics, it was argued that "new schemes of accompaniment, technical supervision, and extensionism" are crucial for community forest development (García García, Marqués Rosano, and Merino 2018, 8; Madrid 2019). However, new schemes may not be necessary, but rather a return to the relatively well-funded policies and programs of the 1970s and 1980s, enriched by the policy experiences of subsequent decades.

In 1937, Simpson noted that

> It is foolish to expect a people for generations chained in the thralldom of feudalism to convert themselves overnight into disciplined, democratic, and peaceful communities. The problem of the political organization and social control of the ejido villages needs to be reexamined by the revolutionary governments. It cannot be solved by laws and paper programs, although both are necessary. . . . [F]or what is involved is the nothing less than the transformation of a large part—in time practically all—of the rural life and institutional make-up of the nation. (352–53)

It was a remarkable historical transformation that indeed so many forest communities converted themselves into disciplined, democratic, and peaceful communities within a largely supportive institutional framework. Over a period of decades, the chaos, disorganization, and rulelessness of the 1930s has been transformed in many hundreds of cases to the evolution of institutions for collective action around the forest commons. In these cases, highly organized communities with clear governance rules are managing their forests for multiple values including biodiversity and generating income for their communities, through the use of active appropriation, provisioning, monitoring, and sanction mechanisms. There are distressingly many cases in which community-level governance remains weak, authoritarian and corrupt community leaders hold power, and deforestation and degradation are present, but the hundreds of successful examples show what is possible when conditions are optimal (Bray 2013). This achievement of reforms in forest governance is still insufficiently recognized in Mexico and beyond and is similar to F. Fukuyama's (2011) characterization of the process of creating democratic political institutions in the West, "the struggle to create modern political institutions was so long and painful that people living in industrialized countries now suffer from a historical amnesia regarding how their societies came to that point in the first place" (14). In a similar sense, there is a historical amnesia about how the Mexican political economy provided an institutional framework that encouraged and rewarded community forest entrepreneurial activity. The institutional framework in Mexico was a strongly facilitating political regime that provided constitutional, collective choice, and operational rules from above that reduced transactions costs and rewarded a particular kind of firm, community forest enterprises (North 1993).

David Barkin noted the successful effort by Mexico to achieve food self-sufficiency in the 1950s and 1960s that has been termed the "Mexican miracle" and was "one of the most successful programs to raise peasant welfare by liberating important forces that would encourage and enable them to increase production; it was also one of the most successful land reforms implemented in the twentieth century in the whole world" (2002, 77). The rise and persistence over decades of hundreds of successful CFEs constitute an ongoing and still little-known second "Mexican miracle." This achievement represents one of the most successful rural and forest policy reforms of the twentieth century in the developing world. Supportive forest policy with infusions of the five capitals, including institutions and organizational models, represents a policy mix that stimulates community collective action and provides a path forward for the future of forest resilience in the developing world and one of the seeds of a good Anthropocene.

REFERENCES

Abardía Moros, F., and C. S. Solano. 1995. "Forestry Communities in Oaxaca: The Struggle for Free Market Access." In *Case Studies of Community-Based Forestry Enterprises in the Americas*, edited by N. Forster, 99–119. Madison: Land Tenure Center and Institute for Environmental Studies, University of Wisconsin-Madison.

Acharaya, K. P. 2002. "Twenty-Four Years of Community Forestry in Nepal." *International Forestry Review* 4(2): 1.

Acheson, J. M. 2003. *Capturing the Commons: Devising Institutions to Manage the Maine Lobster Industry*. Hanover, N.H.: University Press of New England.

Adger, W. N. 2000. "Social and Ecological Resilience: Are They Related?" *Progress in Human Geography* 24(3): 347–64.

Adger, W. Neil, K. Brown, R. Cervigni, and D. Moran. 1995. "Total Economic Value of Forests in Mexico." *Ambio* 24(5): 286–96.

Adger, W. N., K. Brown, and E. L. Tompkins. 2005. "The Political Economy of Cross-Scale Networks in Resource Co-management." *Ecology and Society* 10(2): 9. http://www.ecologyandsociety.org/vol10/iss2/art9.

Agrawal, A. 2001. "Common Property Institutions and Sustainable Governance of Resources." *World Development* 29(10): 1649–72.

Agrawal, A., and A. Chhatre. 2006. "Explaining Success on the Commons: Community Forest Governance in the Indian Himalaya." *World Development* 34(1): 149–66.

Agrawal, A., and C. C. Gibson. 1999. "Enchantment and Disenchantment: The Role of Community in Natural Resource Conservation." *World Development* 27(4): 629–49.

Aguilar, J., S. Madrid, L. Merino, and P. Gutiérrez. 1990. *Memoria Segundo Taller de Análisis de Experiencias Forestales*, edited by G. Alatorre, 124. Mexico City: ERA-SAED, Programa Pasos, FPH.

Aguilar Espinoza, Roque Oscar. 1990. "Organización forestal en México." Thesis for Forest Engineering, Universidad Autónoma de Chapingo.

Aguirre Beltrán, G. 1979. *Regions of Refuge*. Society for Applied Anthropology Monograph Series: Monograph No. 12. Washington, D.C.: Society for Applied Anthropology.

Alatorre Frenk, G. 2000. *La construcción de una cultura gerencial democrática en las empresas forestales comunitarias*. Mexico City: Casa Juan Pablos, Procuraduría Agraria.

Alchian, A. A., and H. Demsetz. 1972. "Production, Information Costs, and Economic Organization." *American Economic Review* 62(5): 777–95.

Alcorn, J. B., and V. M. Toledo. 1998. "Resilient Resource Management in Mexico's Forest Ecosystems: The Contribution of Property Rights." In *Linking Social and Ecological Systems: Management Practices and Social Mechanisms for Building Resilience*, edited by F. Berkes and C. Folke, 216–49. Cambridge: Cambridge University Press.

Allen, C. D., A. K. Macalady, H. Chenchouni, D. Bachelet, N. McDowell, M. Vennetier, ... and P. Gonzalez. 2010. "A Global Overview of Drought and Heat-Induced Tree Mortality Reveals Emerging Climate Change Risks for Forests." *Forest Ecology and Management* 259(4): 660–84.

Aligica, P. D. 2014. *Institutional Diversity and Political Economy: The Ostroms and Beyond*. Oxford: Oxford University Press.

Aligica, P. D., and P. J. Boettke. 2009. *Challenging Institutional Analysis and Development: The Bloomington School*. London: Routledge.

Alix-Garcia, J., A. De Janvry, E. Sadoulet, and J. Manuel. 2009. "Lessons Learned from Mexico's Payment for Environmental Services Program." In *Payment for Environmental Services in Agricultural Landscapes*, edited by L. Lipper, T. Sakuyama, R. Stringer, and D. Zilberman, 163–88. New York: Springer.

Alix-Garcia, J. M., E. N. Shapiro, and K. R. Sims. 2012. "Forest Conservation and Slippage: Evidence from Mexico's National Payments for Ecosystem Services Program." *Land Economics* 88(4): 613–38.

Alix-Garcia, J. M., K. R. Sims, and P. Yañez-Pagans. 2015. "Only One Tree from Each Seed? Environmental Effectiveness and Poverty Alleviation in Mexico's Payment for Ecosystem Services Program." *American Economic Journal: Economic Policy* 7(4): 1–40.

Almazán-Núñez, C. R., F. Puebla-Olivares, and A. Almazán-Juárez. 2009. "Diversidad de aves en bosques de pino-encino del centro de Guerrero, México." *Acta Zoológica Mexicana* (n.s.) 25(1): 123–42.

Alvarez, S., and A. Rubio. 2016. "Wood Use and Forest Management for Carbon Sequestration in Community Forestry in Sierra Juárez, Mexico." *Small-Scale Forestry* 15(3): 357–74.

Amacher, G. S., M. Ollikainen, and E. Koskela. 2009. *Economics of Forest Resources*. Cambridge, Mass.: MIT Press.

AMPF (Asociación Mexicana de Profesionistas Forestales). 1979. "El Plan Forestal Puebla: Una alternativa de solución al viejo problema de la equidad." *México y sus Bosques* 18:1, 13–19.

Andam, K. S., P. J. Ferraro, A. Pfaff, G. A. Sánchez-Azofeifa, and J. A. Robalino. 2008. "Measuring the Effectiveness of Protected Area Networks in Reducing Deforestation." *Proceedings of the National Academy of Sciences* 105(42): 16089–94.

Anderies, J. M., C. Folke, B. Walker, and E. Ostrom. 2013. "Aligning Key Concepts for Global Change Policy: Robustness, Resilience, and Sustainability." *Ecology and Society* 18(2): 8. https://doi.org/10.5751/ES-05178-180208.

Anderies, J. M., J. D. Mathias, and M. A. Janssen. 2019. "Knowledge Infrastructure and Safe Operating Spaces in Social-Ecological Systems." *Proceedings of the National Academy of Sciences*, March 19, 116 (12): 5277–84.

Angelsen, A., with M. Brockhaus, M. Kanninen, E. Sills, W. D. Sunderlin, and S. Wertz-Kanounnikoff, eds. 2009. "Realising REDD+: National Strategy and Policy Options." Bogor, Indonesia: CIFOR. https://www.cifor.org/publications/pdf_files/Books/BAngelsen0902.pdf.

Angulo Carrera, A., and M. Martínez Tapia. 2008. *La guerra por el bosque*. Mexico City: Instituto Politécnico Nacional.

Anta Fonseca, S. 2015. "El fortalecimiento de la silvicultura comunitaria en México: Una perspectiva de atención desde la Conafor." In *Desarrollo forestal comunitario: La política pública*, edited by J. M. Torres, 145–74. Mexico City: Centro de Investigación y Docencia Económica (CIDE).

Anta Fonseca, S., A. V. Arreola Muñoz, M. A. González Ortiz, and J. Acosta González, eds. 2006. *Ordenamiento territorial comunitario: Un debate de la sociedad civil hacia la construcción de políticas públicas*. Mexico City: Instituto Nacional de Ecología (SEMARNAT-INE).

Anta Fonseca, S., and F. Mondragón Galicia. 2006. "El ordenamiento territorial y los estatutos comunales: El caso de Santa Cruz Tepetotutla, Usila, Oaxaca." In *Ordenamiento territorial comunitario: Un debate de la sociedad civil hacia la construcción de políticas públicas*, edited by S. Anta Fonseca, A. V. Arreola Muñoz, M. A. González Ortiz, and J. Acosta González, 191–208. Mexico City: Instituto Nacional de Ecología (SEMARNAT-INE).

Antinori, C. 2000. *Vertical Integration in Mexican Common Property Forests*. PhD diss., Department of Agricultural and Resource Economics, University of California, Berkeley. https://are.berkeley.edu/~cmantinori/cmthesis.pdf.

Antinori, C. 2005. "Vertical Integration in the Community Forest Enterprises of Oaxaca." In *The Community Forests of Mexico: Managing for Sustainable Landscapes*, edited by D. B. Bray, L. Merino-Pérez, and D. Barry, 241–72. Austin: University of Texas Press.

Antinori, C., and D. B. Bray. 2005. "Community Forest Enterprises as Entrepreneurial Firms: Economic and Institutional Perspectives from Mexico." *World Development* 33(9): 1529–43.

Antinori, C., and G. Rausser. 2003. "Does Community Involvement Matter? How Collective Choice Affects Forests in Mexico." January 10. Department of Agricultural and Resource Economics, CUDARE Working Papers 939, University of California, Berkeley. https://escholarship.org/content/qt83j385n0/qt83j385n0.pdf.

Antinori, C., and G. Rausser. 2007. "Collective Choice and Community Forestry Management in Mexico: An Empirical Analysis." *Journal of Development Studies* 43(3): 512–36.

Antinori, C., and G. C. Rausser. 2010. "The Mexican Common Property Forestry Sector." CUDARE Working Papers 1105. Department of Agricultural and Resource Economics. University of California, Berkeley. https://ageconsearch.umn.edu/record/90936/files/CUDARE%201105%20Antinori.pdf.

Armendáriz Payán, H. M. 2014. "Método Mexicano de Ordenación de Bosques Irregulares (MMOBI)." Curso Regional de Regulación, Manejo y Salud Forestal: SEMARNAT. http://biblioteca.semarnat.gob.mx/janium/Documentos/Ciga/Libros2014/CD001808.pdf.

Armijo Canto, N., and J. Robertos Jiménez. 1997. "Distribución de los beneficios socioeconómicos del bosque." Technical report. Chetumal, Mexico: Universidad de Quintana Roo / Departamento para el Desarrollo Internacional (Gobierno Británico).

Arnold, J. E. M. 1998. *Managing Forests as Common Property*. Rome: FAO, 1998.

ASETECO. 2002. *Una caminata de veinte años en los bosques comunales de Oaxaca*. Oaxaca City: ASETECO/CECAMO.

Assies, W. 2008. "Land Tenure and Tenure Regimes in Mexico: An Overview." *Journal of Agrarian Change* 8(1): 33–63.

Aviña, A. 2014. *Specters of Revolution: Peasant Guerrillas in the Cold War Mexican Countryside*. Oxford: Oxford University Press.

Axelrod, R. 1984. *The Evolution of Cooperation*. New York: Penguin Books.

Bailón Corres, J. 1999. *Pueblos indios, élites y territorio*. Mexico City: El Colegio de México.

Barkin, D. 2002. "The Reconstruction of a Modern Mexican Peasantry." *Journal of Peasant Studies* 30(1): 73–90.

Barsimantov, J. A. 2010. "Vicious and Virtuous Cycles and the Role of External Nongovernment Actors in Community Forestry in Oaxaca and Michoacán, Mexico." *Human Ecology* 38(1): 49–63.

Barsimantov, J., and J. Navia Antezana. 2012. "Forest Cover Change and Land Tenure Change in Mexico's Avocado Region: Is Community Forestry Related to Reduced Deforestation for High Value Crops?" *Applied Geography* 32(2): 844–53.

Barsimantov, J., and J. Kendall. 2012. "Community Forestry, Common Property, and Deforestation in Eight Mexican States." *Journal of Environment and Development* 21(4): 414–37.

Bartra, A. 1991. "Pro, contras y asegunes de la 'apropiación del proceso productivo.'" In *Los nuevos sujetos del desarrollo rural*, edited by A. Bartra et al., 3–22. Mexico City: ADN Editores.

Bartra, A. 2000. *Crónicas del sur: Utopías campesinas en Guerrero*. Mexico City: Ediciones Era.

Begon, M., J. L. Harper, and C. R. Townsend. 1986. *Ecology: Individuals, Populations and Communities*. Oxford: Blackwell Scientific Publications.

Benet, R. 2018. "El fracaso de la política forestal de Peña Nieto." *Aristegui Noticias*. November 14. https://aristeguinoticias.com/1411/mexico/el-fracaso-de-la-politica-forestal-de-pena-nieto-articulo/.

Benneker, C. E. B. 2008. *Dealing with the State, the Market and NGOs: The Impact of Institutions on the Constitution and Performance of Community Forest Enterprises (CFE) in the Lowlands of Bolivia.* PhD diss., Wageningen University, Wageningen, the Netherlands. https://library.wur.nl/WebQuery/wurpubs/fulltext/122067.

Bennett, E. M., M. Solan, R. Biggs, T. McPhearson, A. V. Norström, P. Olsson, et al. 2016. "Bright Spots: Seeds of a Good Anthropocene." *Frontiers in Ecology and the Environment* 14(8): 441–48.

Berget, C., E. Durán, and D. B. Bray. 2015. "Participatory Restoration of Degraded Agricultural Areas Invaded by Bracken Fern (*Pteridium aquilinum*) and Conservation in the Chinantla Region, Oaxaca, Mexico." *Human Ecology* 43(4): 547–58.

Berkes, F., D. Feeny, B. J. McCay, and J. M. Acheson. 1989. "The Benefits of the Commons." *Nature* 340(6229): 91.

Berkes, F., and H. Ross. 2013. "Community Resilience: Toward an Integrated Approach." *Society and Natural Resources* 26(1): 5–20.

Blackman, A., A. Raimondi, and F. Cubbage. 2017. "Does Forest Certification in Developing Countries Have Environmental Benefits? Insights from Mexican Corrective Action Requests." *International Forestry Review* 19(3): 247–64.

Blomley, T., and S. Iddi. 2009. Participatory Forest Management in Tanzania: 1993–2009. Lessons learned and experiences to date. Unpublished report. http://typo3.p264412.webspaceconfig.de/fileadmin/_migrated/content_uploads/EXPERIENCE_AND_LESSONS_LEARNED_IN_PARTICIPATORY_FOREST_MANAGEMENT_1_.pdf.

Boege, E. 2008. *El patrimonio biocultural de los pueblos indígenas de México: Hacia la conservación in situ de la biodiversidad y agrodiversidad en los territorios indígenas.* Mexico City: Instituto Nacional de Antropología e Historia / Comisión Nacional para el Desarrollo de los Pueblos Indígenas.

Boillat, S., Gerber, J. D., Oberlack, C., Zaehringer, J., Ifejika Speranza, C., and Rist, S. 2018. "Distant Interactions, Power, and Environmental Justice in Protected Area Governance: A Telecoupling Perspective." *Sustainability* 10(11): 3954.

Borrini-Feyerabend, G., A. Kothari, and G. Oviedo. 2004. *Indigenous and Local Communities and Protected Areas: Towards Equity and Enhanced Conservation.* Cambridge: International Union for Conservation of Nature.

Boyce, J., and B. Shelley, eds. 2003. *Natural Assets: Democratizing Ownership of Nature.* Washington, D.C.: Island Press.

Boyer, C. 2015. *Political Landscapes: Forests, Conservation, and Community in Mexico.* Durham, N.C.: Duke University Press.

Bowles, S., and Gintis, H. 1998. "The Moral Economy of Communities: Structured Populations and the Evolution of Pro-social Norms." *Evolution and Human Behavior* 19(1): 3–25.

Bowles, S., and H. Gintis. 2011. *A Cooperative Species: Human Reciprocity and Its Evolution.* Princeton, N.J.: Princeton University Press.

Bragg, D. C., and J. M. Guldin. 2010. *Integrated Management of Carbon Sequestration and Biomass Utilization Opportunities in a Changing Climate: Proceedings of the 2009*

National Silviculture Workshop; 2009 June 15–18; Boise, ID, technically edited by Theresa B. Jain, Russell T. Graham, and Jonathan Sandquist. Proceedings RMRS-P-61. Fort Collins, Colo.: U.S. Department of Agriculture, Forest Service, Rocky Mountain Research Station.

Brandon, K., L. J. Gorenflo, A. S. Rodrigues, and R. W. Waller. 2005. "Reconciling Biodiversity Conservation, People, Protected Areas, and Agricultural Suitability in Mexico." *World Development* 33(9): 1403–18.

Bray, D. B. 1991. "The Struggle for the Forest: Conservation and Development in the Sierra Juárez." *Grassroots Development, Journal of the Inter-American Foundation* 15(3): 12–25.

Bray, D. B. 1995. "Peasant Organizations and 'the Permanent Reconstruction of Nature': Grassroots Sustainable Development in Rural Mexico." *Journal of Environment and Development* 4 (2): 185–204.

Bray, D. B. 2002. "A Purépecha Community Conserves Its Forests While Creating Wealth." *In Sustainable Solutions: Building Assets for Empowerment and Sustainable Development.* Ford Foundation: New York. http://www.ccmss.org.mx/acervo/a-purepecha-community-conserves-its-forests-while-creating-wealth/.

Bray, D. B. 2004a. "Manejo adaptativo, organizaciones y manejo de la propiedad común: Perspectivas de los bosques comunales de Quintana Roo, México." In *Uso, conservación y cambio en los bosques de Quintana Roo,* edited by N. Armijo and C. Llorens, 56–85. Mexico City: Universidad de Quintana Roo / ONACYT.

Bray, D. B. 2004b. "Community Forestry as a Strategy for Sustainable Management: Perspectives from Quintana Roo, Mexico." In *Working Forests in the American Tropics: Conservation Through Sustainable Management,* edited by Daniel Zarin, Janaki Alavalapati, Francis E. Putz, and Marianne C. Schmink, 221–37. New York: Columbia University Press.

Bray, D. B. 2007. "A Path Through the Woods: Community Forest Management in Mexico." *Grassroots Development: Journal of the Inter-American Foundation* 28(1): 40–47. https://archive.iaf.gov/resources/publications/grassroots-development-journal/2007-african-descendants-and-development/a-path-through-the-woods-community-forest-management-in-mexico.html.

Bray, D. B. 2008. "Collective Action, Common Property Forests, Communities and Markets." *Commons Digest* 6:1–4. http://dlc.dlib.indiana.edu/dlc/bitstream/handle/10535/5942/commons%20forum%20commentary.pdf?sequence=1&isAllowed=y.

Bray, D. B. 2010a. "Towards 'Post-REDD Landscapes': Mexican Community Forest Enterprises Provide a Proven Pathway to Reduce Emissions from Deforestation and Forest Degradation." *Infobrief* No. 30. Bogor, Indonesia: Center for International Forestry Research (CIFOR).

Bray, D. B. 2010b. "Forest Cover Dynamics and Forest Transitions in Mexico and Central America: Towards a Great Restoration"? In *Reforesting Landscapes: Linking Pattern and Process,* edited by H. Nagendra and J. Southworth, 85–120. New York: Springer.

Bray, D. B. 2012. "Carbon and Community Development: An Experiment in Oaxaca." *Grassroots Development: Journal of the Inter-American Foundation* 28(1): 15–21.

Bray, D. B. 2013. "When the State Supplies the Commons: Origins, Changes, and Design of Mexico's Common Property Regime." *Journal of Latin American Geography* 12(1): 33–55.

Bray, D. B. 2016. "Muir and Pinchot in the Sierra Norte of Oaxaca: Governance of Forest Management and Forest Recovery in Pueblos Mancomunados." *World Development Perspectives* 4(December): 8–10.

Bray, D. B., C. Antinori, and J. M. Torres-Rojo. 2006. "The Mexican Model of Community Forest Management: The Role of Agrarian Policy, Forest Policy and Entrepreneurial Organization." *Forest Policy and Economics* 8(4): 470–84.

Bray, D. B, M. Carreon, L. Merino, and V. Santos. 1993. "On the Road to Sustainable Forestry: The Maya of Quintana Roo." In *Cultural Survival Quarterly* 17(March): 1.

Bray, D. B., and E. Duran-Medina. 2014. "Options for Reducing Carbon Emissions in Forest Management in the Oaxaca and Chihuahua Áreas de Acción Temprana REDD+." Report for the Nature Conservancy. August. https://www.researchgate .net/publication/326606547_Options_for_Reducing_Carbon_Emissions_in_Forest _Management_in_the_Oaxaca_and_Chihuahua_Areas_de_Accion_Temprana _REDD_AATR_ALIANZA_MEXICO_PARA_LA_REDUCCION_DE _EMISIONES_POR_DEFORESTACION_Y_DEGRADACION.

Bray, D. B., E. Durán, J. Hernández-Salas, C. Luján-Alvarez, M. Olivas-García, and I. Grijalva-Martínez. 2016. "Back to the Future: The Persistence of Horse Skidding in Large Scale Industrial Community Forests in Chihuahua, Mexico." *Forests* 7(11): 283.

Bray, D. B., E. Durán Medina, L. Merino Pérez, J. M. Torres Rojo, and A. Velázquez Montes. 2007. *Nueva evidencia: Los bosques comunitarios de México protegen el ambiente, disminuyen la pobreza y promuevan la paz social.* https://www.ccmss.org.mx/wp -content/uploads/2014/10/Nueva_evidencia_Los_bosques_comunitarios_de_Mexico _protegen_el_ambiente_disminuyen_la_pobreza_y_promueven_paz_social.pdf.

Bray, D. B., E. Durán, and O. Molina-González. 2012. "Beyond Harvests in the Commons: Multi-scale Governance and Indigenous/Community Conserved Areas in Oaxaca, Mexico." *International Journal of the Commons* 6(2): 151–78.

Bray, D. B., E. Duran, V. H. Ramos, J.-F. Mas, A. Velazquez, R. B. McNab, D. Barry, and J. Radachowsky. 2008. "Tropical Deforestation, Community Forests, and Protected Areas in the Maya Forest." *Ecology and Society* 13(2): 56. http://www.ecologyandsociety .org/vol13/iss2/art56/.

Bray, D. B., E. A. Ellis, N. Armijo-Canto, and C. T. Beck. 2004. "The Institutional Drivers of Sustainable Landscapes: A Case Study of the 'Mayan Zone' in Quintana Roo, Mexico." *Land Use Policy* 21(4): 333–46.

Bray, D. B., and P. Klepeis. 2005. "Deforestation and Forest Transitions in Southeastern Mexico, 1900–2000." *Environment and History* 11:195–223.

Bray, D. B, and L. Merino-Pérez. 2002. *The Rise of Community Forestry in Mexico: History, Concepts, and Lessons Learned from Twenty-Five Years of Community Timber Production.* September. A Report to the Ford Foundation, Mexico City. https://www.ccmss .org.mx/wp-content/uploads/2014/09/the_rise_of_community_forestry_in_mexico .pdf.

Bray, D. B., and L. Merino-Pérez. 2003. "El Balcón, Guerrero: A Case Study of Globalization Benefiting a Forest Community." In *Confronting Globalization: Economic Integration and Popular Response in Mexico*, edited by T. A. Wise, H. Salazar, and L. Carlsen, n.p. Bloomfield, Conn.: Kumarian Press.

Bray, D. B., and L. Merino-Pérez. 2004. *La experiencia de las comunidades forestales en México*. Mexico City: Semarnat, INE y CCMSS. http://www.ccmss.org.mx/acervo/la-experiencia-de-las-comunidades-forestales-en-mexico/.

Bray, D. B., L. Merino-Pérez, and D. Barry, eds. 2005a. *The Community Forests of Mexico: Managing for Sustainable Landscapes*. Austin: University of Texas Press.

Bray, D. B., L. Merino-Pérez, and D. Barry. 2005b. "Community Managed in the Strong Sense of the Phrase." In *The Community Forests of Mexico: Managing for Sustainable Landscapes*, edited by D. B. Bray, L. Merino-Pérez, and D. Barry, 3–26. Austin: University of Texas Press.

Bray, D. B., L. Merino-Pérez, P. Negreros-Castillo, G. Segura-Warnholtz, J. M. Torres-Rojo, and H. F. Vester. 2003. "Mexico's Community-Managed Forests as a Global Model for Sustainable Landscapes." *Conservation Biology* 17(3): 672–77.

Breach Velducea, M. 2016. "Devasta el crimen bosques de la Tarahumara." *La Jornada*. June 18. https://www.jornada.com.mx/2016/06/18/contraportada.pdf.

Briones-Salas M., A. Hernández-Allende, M. Martínez, and G. González. 2012. "New Records of the Endemic Chinanteco Deermouse *Habromys chinanteco* (Rodentia: Cricetidae) in the Sierra Madre de Oaxaca, Mexico." *Southwestern Naturalist* 57(2): 222–23.

Briones-Salas, M., and V. Sánchez-Cordero. 2004. "Mamíferos." *Biodiversidad de Oaxaca*, edited by A. García-Mendoza, M. J. Ordóñez, and M. Briones-Salas, 423–48. Mexico City: Instituto de Biología UNAM / Fondo Oaxaqueño para la Conservación de la Naturaleza / WWF.

Brower, L. P., G. Castilleja, A. Peralta, J. López-García, L. Bojorquez-Tapia, S. Díaz, . . . and M. Missrie. 2002. "Quantitative Changes in Forest Quality in a Principal Overwintering Area of the Monarch Butterfly in Mexico, 1971–1999." *Conservation Biology* 16(2): 346–59.

Bruggemann, J. H., M. Rodier, M. M. Guillaume, S. Andréfouët, R. Arfi, J. E. Cinner . . . and T. R. McClanahan. 2012. "Wicked Social-Ecological Problems Forcing Unprecedented Change on the Latitudinal Margins of Coral Reefs: The Case of Southwest Madagascar." *Ecology and Society* 17(4): 47.

Buendía, E. 2016. "Ubican 108 zonas críticas de deforestación en el país." *El Universal*. March 20. https://www.eluniversal.com.mx/articulo/periodismo-de-datos/2016/03/20/ubican-108-zonas-criticas-de-deforestacion-en-el-pais.

Burivalova, Z., Ç. H. Şekercioğlu, and L. P. Koh. 2014. "Thresholds of Logging Intensity to Maintain Tropical Forest Biodiversity." *Current Biology* 24(16): 1893–98.

Burkett, V. R., D. A. Wilcox, R. Stottlemyer, W. Barrow, D. Fagre, J. Baron . . . and G. Ruggerone. 2005. "Nonlinear Dynamics in Ecosystem Response to Climatic Change: Case Studies and Policy Implications." *Ecological Complexity* 2(4): 357–94.

Bustamante Álvarez, T. 1996. "Los recursos forestales de Guerrero, su aprovechamiento social y la apertura comercial: El caso del ejido el balcón." In *El acceso a los recursos naturales y el desarrollo sustentable.* Vol. 3, *La sociedad rural mexicana frente al nuevo milenio,* edited by H. Mackinlay and E. Boege, 367–84. Mexico City: Plaza y Valdés Editores.

Caballero Deloya, Miguel. 2008. "El 'otro' México forestal (La actividad forestal ilícita)." *Ciencia Forestal en México* 33(103): 149–75.

Caballero Deloya, Miguel. 2010. "La verdadera cosecha maderable en México." *Revista Mexicana de Ciencias Forestales* 1(1): 5–16.

Cairns, M. A., P. K. Haggerty, R. Alvarez, B. H. De Jong, and I. Olmsted. 2000. "Tropical Mexico's Recent Land-Use Change: A Region's Contribution to the Global Carbon Cycle." *Ecological Applications* 10(5): 1426–41.

Calva, J. L. Téllez, F. Paz González, O. Wicab Gutiérrez, and J. Camas Reyes. 1989. *Economía política de la explotación forestal en México: Bibliografía comentada 1930–1984.* 1st ed. Mexico City: Universidad Autónoma Chapingo, Universidad Nacional Autónoma de México.

Canseco-Márquez, L., C. G. Ramírez-González, and E. Gonzalez-Bernal. 2017. "Discovery of Another New Species of Charadrahyla (Anura, Hylidae) from the Cloud Forest of Northern Oaxaca, México." *Zootaxa* 4329(1): 64–72.

Carillo-Anzures, F. C., M. A. Mireles, E. F. Ayala, J. M. Torres Rojo, D. M. Sangerman-Jarquín, L. G. Molina, and E. B. Rodríguez. 2017. "Caracterización de productores forestales en 12 estados de la República Mexicana." *Revista Mexicana de Ciencias Agrícolas* 8(7): 1561–73.

Carley, M., and I. Christie. 2000. *Managing Sustainable Development.* 2nd ed. Minneapolis: University of Minnesota Press.

Carmagnani, M. 1993. *El regreso de los dioses: El proceso de reconstitución de la identidad étnica en Oaxaca. Siglos XVII y XVIII.* Mexico City: Fondo de Cultura Económica.

Carroll, M., B. Milakovsky, A. Finkral, A. Evans, and M. S. Ashton. 2012. "Managing Carbon Sequestration and Storage in Temperate and Boreal Forests." In *Managing Forest Carbon in a Changing Climate,* edited by M. A. Ashton, M. L. Tyrrel, D. Spalding, and B. Gentry, 205–26. New York: Springer.

Cashore, B., G. Auld, and D. Newsom. 2004. *Governing Through Markets: Forest Certification and the Emergence of Non-state Authority.* New Haven, Conn.: Yale University Press.

Castaños Martínez, L. J. 2015. "Esfuerzos pioneros de política pública para la gestión forestal en comunidades: La Dirección General para el Desarrollo Forestal (1973–1985)." In *Desarrollo forestal comunitario,* edited by J. M. Torres Rojo, 39–70. Mexico City: Centro de Investigación y Docencia Económica.

Cayuela, L., J. M. R. Benayas, and C. Echeverría. 2006. "Clearance and Fragmentation of Tropical Montane Forests in the Highlands of Chiapas, Mexico (1975–2000)." *Forest Ecology and Management* 226(1–3): 208–18.

CCMSS. 2006. "Balance de la orientación de la política forestal nacional al final de la presente administración." *Nota Informativa Número 11.* October. Mexico City: Consejo Civil Mexicano para la Silvicultura Sostenible. https://www.ccmss.org.mx/acervo/nota-informativa-11-balance-orientacion-politica-forestal-nacional-al-final-de-la-presente-administracion/.

CCMSS. 2007. "El combate a la tala ilegal en México." *Nota Informativo Número 15.* June. Mexico City: Consejo Civil Mexicano para la Silvicultura Sostenible. https://www .ccmss.org.mx/acervo/nota-informativa-15-el-combate-a-la-tala-ilegal-en-mexico/.

CCMSS. 2008a. "Tramitología: Un fuerte obstáculo para el sector forestal en México." *Nota Informativo Número 19.* June. Consejo Civil Mexicano para la Silvicultura Sostenible. CMSS: Mexico City. http://www.ccmss.org.mx/wp-content/uploads/2014/10/nota_info_19_tramitologia.pdf.

CCMSS. 2008b. ProÁrbol 2008: "Cambios presupuestales y de operación." *Nota Informativo Número 17.* February. Mexico City: Consejo Civil Mexicano para la Silvicultura Sostenible. https://www.ccmss.org.mx/acervo/nota-informativa-17-proarbol -2008-cambios-presupuestales-y-de-operacion/.

CCMSS. 2010. "Análisis del Proyecto de Presupuesto para CONAFOR en 2011." *Nota Informativo 29.* September. Mexico City: Consejo Civil Mexicano para la Silvicultura Sostenible. https://www.ccmss.org.mx/acervo/nota-informativa-29-presupuesto -forestal-2011/.

CCMSS. 2012. "Un nuevo enfoque para combatir la tala y el comercio de madera ilegal en México." *Nota Informativo Número 33.* Mexico City: Consejo Civil Mexicano para la Silvicultura Sostenible. CMSS. https://www.ccmss.org.mx/acervo/nota-informativa -33-un-nuevo-enfoque-para-combatir-la-tala-y-el-comercio-de-madera-ilegal-en -mexico/.

CCMSS. 2014. "Presupuesto forestal 2015: Un ejercicio de opacidad." *Nota Informativo 39.* October. Mexico City: Consejo Civil Mexicano para la Silvicultura Sostenible. https://www.ccmss.org.mx/acervo/nota-informativa-39-presupuesto-forestal-2015 -un-ejercicio-de-opacidad/.

CCMSS. 2016. "Presupuesto forestal 2016: ¿Nuevos riesgos para los bosques?" *Nota Informativa 43.* October. Mexico City: Consejo Civil Mexicano para la Silvicultura Sostenible. https://www.ccmss.org.mx/acervo/nota-informativa-43-presupuesto -forestal-2016-nuevos-riesgos-para-los-bosques/.

Ceballos, G., P. Rodríguez, and R. Medellín. 1998. "Assessing Conservation Priorities in Megadiverse Mexico: Mammalian Diversity, Endemicity, and Endangerment." *Ecological Applications* 8(1): 8–17.

Ceballos, G., C. Chávez, H. Zarza, and C. Manterola. 2005. "Ecología y conservación del jaguar en la región de Calakmul." *Biodiversitas* 62(September): 1–7.

CFR-LA. 2016. *Promise and Performance: Ten Years of the Forest Rights Act in India. Citizens' Report on Promise and Performance of the Scheduled Tribes and Other Traditional Forest Dwellers (Recognition of Forest Rights) Act, 2006, After 10 Years of Its Enactment.* December. Produced as part of Community Forest Rights-Learning and Advocacy Process (CFRLA), India.

Ciriacy-Wantrup, S. V., and R. C. Bishop. 1975. "'Common Property' as a Concept in Natural Resources Policy." *Natural Resources Journal* 15(4): 713–27.

Challenger, A. 1998. *Utilización y conservación de los ecosistemas terrestres de México: Pasado, presente futuro.* Mexico City: Comisión Nacional para el Conocimiento y Uso de la Biodiversidad.

Challenger, A., and J. Soberón. 2008. "Los ecosistemas terrestres." *Capital Natural de México.* Vol. 1, *Conocimiento Actual de la Biodiversidad*, compiled by J. Soberón, G. Halffter, and J. Llorente-Bousquets, 87–108. Mexico City: Comisión Nacional para el Conocimiento y Uso de la Biodiversidad.

Chambille, K. 1983. *Atenquique: Los bosques del sur de Jalisco.* Mexico City: Instituto de Investigaciones Económicas, Universidad Nacional Autónoma de México.

Chapela y Mendoza, F. 2006. "Reconocimiento de los derechos colectivos y Ordenamiento de los Territorios Comunales en América Latina y el Caribe." In *Ordenamiento Territorial Comunitario: Un debate de la sociedad civil hacia la construcción de políticas públicas*, edited by S. Anta Fonseca, A. V. Arreola Muñoz, M. A. González Ortiz, and J. Acosta González, 53–68. Mexico City: Instituto Nacional de Ecología (SEMARNAT-INE).

Chapela y Mendoza, F., and Y. Lara. 2007. "El Ordenamiento Comunitario del Territorio: Un esquema para hacer compatibles los objetivos de conservación y los derechos sociales e indígenas." *Policy Matters* 15:288–98.

Chapela y Mendoza, G. 1996. "La política de aprovechamiento forestal en México: Liberalismo, comunidades y conservación de bosque." In *El acceso a los recursos naturales y el desarrollo sustentable.* Vol. 3, *La sociedad rural mexicana frente al nuevo milenio*, edited by H. Mackinlay and Boege E, 342–65. Mexico City: Plaza y Valdés Editores.

Chapela y Mendoza, G. 1998. "La organización campesina forestal en el cambio liberal Mexicano: 1980–1992." PhD diss., Universidad Nacional Autónoma de México, Mexico City.

Chapela y Mendoza, G. 2012. "Problemas y oportunidades en el Mercado para las empresas sociales forestales en México." Mexico City: Consejo Civil Mexicano para la Silvicultura Sostenible. https://www.ccmss.org.mx/acervo/problemas-y-oportunidades -en-el-mercado-para-las-empresas-sociales-forestales-en-mexico-borrador/.

Chapela y Mendoza, G., and R. Meraz. n.d. *Los grandes ejidos de Durango.* Manuscript available from author.

Chapela, G. 2018. "Ley General de Desarrollo Forestal Sustentable: Un balance favorable con claroscuros." April. Manuscript. https://ceiba.org.mx/publicaciones/Leyes& Normas/180428_LGDFS_Balance_GChapela.pdf.

Chapela y Mendoza, G. 2018a. "Los bosques de México: Crisis del sector forestal y propuesta de política." In *Agenda Ambiental: 2018*, edited by L. Merino Pérez and A. Velázquez, 33–38. Accessed November 1, 2018. https://www.researchgate.net/ publication/324703249_Agenda_Ambiental_2018_Diagnostico_y_propuestas.

Chapela y Mendoza, G., 2018b. "Avances y retrocesos en la integración productiva de las comunidades forestales de México." *Las empresas sociales forestales en México: Claroscuros y aprendizajes.* Mexico City: Consejo Civil Mexicano para la Silvicultura Sostenible, AC. Accessed December 10, 2019. www.ccmss.org.mx/acervo/las -empresas-sociales-forestales-en-mexico-claroscuros-y-aprendizajes/.

Chhatre, A., and A. Agrawal. 2008. Forest Commons and Local Enforcement. *Proceedings of the National Academy of Sciences* 105(36): 13286–291.

Cleaver, F., and J. De Koning. 2015. "Furthering Critical Institutionalism." *International Journal of the Commons* 9(1): 1–18.

Chizcagum. 2020. Advertisement. Accessed March 14, 2020. http://www.chiczagum.ca/home/.

Collins, S. L., S. R. Carpenter, S. M. Swinton, D. E. Orenstein, D. L. Childers, T. L. Gragson, . . . and A. K. Knapp. 2011. "An Integrated Conceptual Framework for Long-Term Social–Ecological research." *Frontiers in Ecology and the Environment* 9(6): 351357.

CONAFOR. 2010. *Visión de México sobre Redd+: Hacia una estrategia nacional.* Comisión Nacional Forestal: Zapopan, Jalisco. Accessed August 22, 2016. http://www.conafor .gob.mx:8080/documentos/docs/7/1393Visi%C3%B3n%20de%20M%C3%A9xico %20sobre%20REDD_.pdf.

CONAFOR. 2012a. *Logros y perspectivas del desarrollo forestal en México 2007–2012.* Zapopan, Mexico: Comisión Nacional Forestal.

CONAFOR. 2012b. *Informe de rendición de cuentas de la Administración Pública Federal 2006–2012. Informe consolidado.* December 1, 2006–November 30, 2012. https://www.gob.mx/cms/uploads/attachment/file/126550/Informes_Gestiones_Anteriores _-_CONAFOR_2006-2012_consolidado.pdf.

CONAFOR. 2014. *Programa Nacional Forestal 2014–2018.* Guadalajara: CONAFOR.

"Consulta temática." 2015. SEMARAT. Accessed August 21, 2016. http://dgeiawf .semarnat.gob.mx:8080/ibi_apps/WFServlet?IBIF_ex=D3_RFORESTA01_08& IBIC_user=dgeia_mce&IBIC_pass=dgeia_mce.

Corral-Rivas, J. S., J. M. Torres-Rojo, J. E. Luján Soto, M. G. Nava Miranda, O. A. Aguirre Calderón, and K. V. Gadow. 2016. "Density and Production in the Natural Forests of Durango/Mexico." *Allgemeine Forst- und Jagdzeitung* 187(5–6): 93–103.

Coase, R. H. 1937. "The Nature of the Firm." *Economica* 4(16): 386–405.

Comisión Nacional de Áreas Naturales Protegidas. 2019. "Listado de ADVC." March 7. https://advc.conanp.gob.mx/listado-de-advc/.

Cornejo, M. K. 2004. *Promoting Community Ecotourism Enterprises in Common Property Regimes: A Stakeholder Analysis and Geographic Information Systems Application in Ejido X-Maben in Central Quintana Roo, Mexico.* MA thesis, Florida International University. https://digitalcommons.fiu.edu/cgi/viewcontent.cgi?referer=https://scholar .google.com.mx/&httpsredir=1&article=3827&context=etd.

Costanza, R., R. de Groot, P. Sutton, S. van der Ploeg, S. J. Anderson, I. Kubiszewski, S. Farber, and R. K. Turner. 2014. "Changes in the Global Value of Ecosystem Services." *Global Environmental Change* 26(May): 152–58.

Cox, M. 2014. "Understanding Large Social-Ecological Systems: Introducing the SES-MAD Project." *International Journal of the Commons* 8(2): 265–76.

Cox, M., G. Arnold, and S. Villamayor Tomás. 2010. "A Review of Design Principles for Community-Based Natural Resource Management." *Ecology and Society* 15(4): 38. http://www.ecologyandsociety.org/vol15/iss4/art38/.

Cramaussel, C. 2013. "El sistema de cargos en San Bernardino de Milpillas Chico, Durango." *Culturales* (Mexicali) 1(1): 69–86.

Cronkleton, P., D. B. Bray, and G. Medina. 2011. "Community Forest Management and the Emergence of Multi-scale Governance Institutions: Lessons for REDD+ Development from Mexico, Brazil and Bolivia." *Forests* 2(2): 451–73.

Cubbage, F., R. Davis, D. Rodríguez Paredes, G. Frey, R. Mollenhauer, Y. Kraus Eclsin, I. A. González Hernández, H. Albarrán Hurtado, A. M. Salazar Cruz, D. N. Chemor Salas. 2013. *Competitividad y acceso a mercados de empresas forestales comunitarias en México.* June. PROFOR / CONAFOR / World Bank. https://www.profor.info/sites/ profor.info/files/MX%20PROFOR-%20Competitiveness%20Report.pdf.

Cubbage, F. W., R. R. Davis, D. Rodríguez Paredes, R. Mollenhauer, Y. Kraus Elsin, G. E. Frey . . . and D. N. C. Salas. 2015. "Community Forestry Enterprises in Mexico: Sustainability and Competitiveness." *Journal of Sustainable Forestry* 34(6–7): 623–50.

Dachary, A. C., and S. M. Arnaiz B. 1983. *Estudios socioeconómicos preliminares de Quintana Roo: Sector agropecuario y forestal (1902–1980).* Puerto Morelos: Centro de Investigaciones de Quintana Roo.

Dalle, S. P., S. de Blois, J. Caballero, and T. Johns. 2006. "Integrating Analyses of Local Land-Use Regulations, Cultural Perceptions and Land-Use / Land Cover Data for Assessing the Success of Community-Based Conservation." *Forest Ecology and Management* 222(1–3): 370–83.

Dalle, S. P., M. T. Pulido, and S. D. Blois. 2011. "Balancing Shifting Cultivation and Forest Conservation: Lessons from a 'Sustainable Landscape' in Southeastern Mexico." *Ecological Applications* 21(5): 1557–72.

Dearing, J. A., R. Wang, K. Zhang, J. G. Dyke, H. Haberl, M. S. Hossain . . . and J. Carstensen. 2014. "Safe and Just Operating Spaces for Regional Social-Ecological Systems." *Global Environmental Change* 28 (September): 227–38.

Dumond, D. E. 1997. *The Machete and the Cross: Campesino Rebellion in Yucatan.* Lincoln: University of Nebraska Press.

de Ita, Ana. 1996. "Política forestal: Entre el bosque natural y las plantaciones forestales comerciales." *Cuadernos Agrarios* 6(14): 83–98.

de Janvry, A., C. Dutilly, C. Muñoz-Piña, and E. Sadoulet. 2001. "Liberal Reforms and Community Responses in Mexico." In *Communities and Markets in Economic Development*, edited by M. Aoki and Y. Hayami, 318–43. Oxford: Oxford University Press.

Deschamps Ramírez, P., and S. Madrid Zubirán. 2018. *Subsidios forestales sin rumbo: Apuntes para una política en favor de las comunidades y sus bosques.* Mexico City: Consejo Civil Mexicano para la Silvicultura Sostenible. http://www.ccmss.org.mx/ acervo/subsidios-forestales-sin-rumbo-apuntes-para-una-politica-en-favor-de-las -comunidades-y-sus-bosques/.

Deschamps Ramírez, P., B. Zavariz Romero, I Zúñiga Pérez-Tejada. 2015. *Revisión de la implementación de REDD+ en México: Análisis de los programas especiales en áreas de Acción Temprana REDD+.* Mexico City: Consejo Civil Mexicano para la Silvicultura Sostenible. http://www.ccmss.org.mx/acervo/revision-de-la-implementacion-de -redd-en-mexico/.

de Tocqueville, A. [1835] 2002. *Democracy in America.* New York: Bantam Classic.

de Vos, J. 1996. *Oro verde: La conquista de la Selva Lacandona por los madereros tabasqueños, 1822–1949.* Villahermosa: Gobierno del Estado de Tabasco, Instituto de Cultura de Tabasco.

Dietz, T., E. Ostrom, and P. C. Stern. 2003. "The Struggle to Govern the Commons." *Science* 302(5652): 1907–12.

Diffenbaugh, N. S., and C. B. Field. 2013. "Changes in Ecologically Critical Terrestrial Climate Conditions." *Science* 341(6145): 486–92.

DiGiano, M., E. Ellis, and E. Keys. 2013. "Changing Landscapes for Forest Commons: Linking Land Tenure with Forest Cover Change Following Mexico's 1992 Agrarian Counter-reforms." *Human Ecology* 41(5): 707–23.

Donangelo, R., H. Fort, V. Dakos, M. Scheffer, and E. H. Van Nes. 2010. "Early Warnings for Catastrophic Shifts in Ecosystems: Comparison Between Spatial and Temporal Indicators." *International Journal of Bifurcation and Chaos* 20(02): 315–21.

Durán, E., D. B. Bray, A. Velázquez, and A. Larrazábal. 2011. "Multi-scale Forest Governance, Deforestation, and Violence in Two Regions of Guerrero, Mexico." *World Development* 39(4): 611–19.

Durán, E. J.-F. Mas, and A. Velasquez. 2005. "Land Use / Land Cover Change in Community-Based Forest Management Regions and Protected Areas in Mexico." In *The Community Forests of Mexico: Managing for Sustainable Landscapes,* edited by D. B. Bray, L. Merino Pérez, and D. Barry, 215–40. Austin: University of Texas Press.

Durán, E., and A. Poloni. 2014. "Escarabajos descortezadores: Diversidad y saneamiento en bosques de Oaxaca." *Biodiversitas* 117:7–12.

Eakin, H. C., M. C. Lemos, and D. R. Nelson. 2014. "Differentiating Capacities as a Means to Sustainable Climate Change Adaptation." *Global Environmental Change* 27(July): 1–8.

Eguiluz-Piedra, T. 2004. "The Present Situation of Mexican Forestry: Food and Agriculture Organization (FAO:Rome)." http://www.fao.org/3/Y4829E/y4829e09.htm.

Ellis, E. A., K. A. Kainer, J. A. Sierra-Huelsz, P. Negreros-Castillo, D. Rodriguez-Ward, and M. DiGiano. 2015. "Endurance and Adaptation of Community Forest Management in Quintana Roo, Mexico." *Forests* 6(11): 4295–327.

Ellis, E. A., S. A. Montero, I. U. H. Gómez, J. A. R. Montero, P. W. Ellis, D. Rodríguez-Ward, . . . and F. E. Putz. 2019. "Reduced-Impact Logging Practices Reduce Forest Disturbance and Carbon Emissions in Community Managed Forests on the Yucatán Peninsula, Mexico." *Forest Ecology and Management* 437(April): 396–410.

Ellis, E. A., and L. Porter-Bolland. 2008. "Is Community-Based Forest Management More Effective than Protected Areas?: A Comparison of Land Use / Land Cover Change in Two Neighboring Study Areas of the Central Yucatan Peninsula, Mexico." *Forest Ecology and Management* 256(11): 1971–83.

Ellis, E. A., J. A. Romero Montero, and I. U. Hernández Gómez. 2017. "Deforestation Processes in the State of Quintana Roo, Mexico: The Role of Land Use and Community Forestry." *Tropical Conservation Science* 10 (April 7). https://doi.org/10.1177/1940082917697259.

Ellis, E. C. 2018. *Anthropocene: A Very Short Introduction*. Oxford: Oxford University Press.

Ellis, P. W., T. Gopalakrishna, R. C. Goodman, F. E. Putz, A. Roopsind, P. M. Umunay . . . and B. W. Griscom. 2019. "Reduced-Impact Logging for Climate Change Mitigation (RIL-C) Can Halve Selective Logging Emissions from Tropical Forests." *Forest Ecology and Management* 438(April): 255–66.

Enciso, A. 2016. "Participará gendarmería en 108 zonas críticas forestales." *La Jornada en Línea*. September 21. https://www.zocalo.com.mx/new_site/articulo/participara -gendarmeria-en-108-zonas-criticas-forestales-1474482538.

Enríquez Quintana, M. 1976. "Las empresas ejidales forestales." *Revista del México Agrario* 9(2): 71–98.

ERA/UZACHI. 1994. *Planeación comunitaria y manejo forestal sostenible*. Oaxaca: ERA/ UZACHI.

Espinosa, D. S., and S. Ocegueda, et al. 2008. "El conocimiento biogeográfico de las especies y su regionalización natural." In *Capital natural de México*. Vol. 1, *Conocimiento actual de la biodiversidad*, compiled by J. Soberón, G. Halffter, and J. Llorente-Bousquets, 33–66. Mexico City: Comisión Nacional para la Biodiversidad.

Estes, J. A., J. Terborgh, J. S. Brashares, M. E. Power, J. Berger, W. J. Bond . . . and R. J. Marquis. 2011. "Trophic Downgrading of Planet Earth." *Science* 333(6040): 301–6.

Estrada, O. 2018. "Diagnóstico del manejo forestal comunitario en Chihuahua." In *Las empresas sociales forestales en México: Claroscuros y aprendizajes*, edited by G. Chapela, 41–58. Mexico City: Consejo Civil Mexicano para la Silvicultura Sostenible, AC.

Estrada, J., R. Villalpando, and E. Olivares. 2018. "Asesinan a defensor de bosques en Chihuahua." *La Jornada*. October 26. https://www.jornada.com.mx/2018/10/26/ estados/032n1est#.

Fama, E. F., and M. C. Jensen. 1983. "Separation of Ownership and Control." *Journal of Law and Economics* 26(2): 301–25.

FAOSTAT. n.d. "FORESTstat." Food and Agricultural Organization. Accessed October 5, 2018. http://www.fao.org/faostat/en/#data/FO.

Farjon, A., and B. T. Styles. 1997. *Pinus (Pinaceae)*. Flora Neotropica Monograph 75. New York: New York Botanical Garden.

Fedele, G., C. I. Donatti, C. A. Harvey, L. Hannah, and D. G. Hole, D. G. 2019. "Transformative Adaptation to Climate Change for Sustainable Social-Ecological Systems." *Environmental Science and Policy* 101:116–25.

Fernández Vázquez, E., and N. Mendoza Fuente. 2015. *Sobrerregulación forestal: Un obstáculo para el desarrollo sustentable de México*. Mexico City: Consejo Civil Mexican para la Silvicultura Sostenible. Accessed July 18, 2016. http://www.ccmss.org.mx/wp -content/uploads/2015/05/Sobrerregulacion-Mendoza-FernandezVazquez-CCMSS .pdf.

Figel, J. J., E. Durán, and D. B. Bray. 2011. "Conservation of the Jaguar *Panthera onca* in a Community-Dominated Landscape in Montane Forests in Oaxaca, Mexico." *Oryx* 45(4): 554–60.

Figueroa, F., and V. Sánchez-Cordero. 2008. "Effectiveness of Natural Protected Areas to Prevent Land Use and Land Cover Change in Mexico." *Biodiversity Conservation* 17, no. 3223: n.p.

Flachsenberg, H., and H. A. Galletti. 1999. "El manejo forestal de la selva en Quintana Roo, México." In *La selva maya: Conservación y desarrollo*, edited by R. B. Primack, D. B. Bray, H. A. Galletti, and I. Ponciano, 66–84. Mexico City: Siglo XXI.

Fleischman, F. D., N. C. Ban, L. S. Evans, G. Epstein, G. Garcia-Lopez, and S. Villamayor-Tomas. 2014. "Governing Large-Scale Social-Ecological Systems: Lessons from Five Cases." *International Journal of the Commons* 8(2): 428–56.

Folke, C. 2006. "Resilience: The Emergence of a Perspective for Social-Ecological Systems Analyses." *Global Environmental Change* 16(3): 253–67.

Forster, R., L. A. Arguelles, S. Kaatz, and N. Aguilar. 2004. "Market Options and Barriers for Community Produced Timber and Sawnwood from Michoacán, Oaxaca, Guerrero, Campeche and Quintana Roo." *Forest Trends*, University of Quintana Roo, Tropical Rural Latinoamericana, and National Forest Commission, Mexico. https://www.forest-trends.org/wp-content/uploads/imported/mexico-market-study_final_9-19-05-pdf.pdf.

Frechette, A., C. Ginsburg, and W. Walker. 2018. "A Global Baseline of Carbon Storage in Collective Lands. Rights and Resources Initiative." Washington, D.C. https://rightsandresources.org/wp-content/uploads/2016/10/Toward-a-Global-Baseline-of-Carbon-Storage-in-Collective-Lands-November-2016-RRI-WHRC-WRI-report.pdf.

Friedrich, P. 1970. *Agrarian Revolt in a Mexican Village*. Englewood Cliffs, N.J.: Prentice-Hall.

Fox, J. 1992. *The Politics of Food in Mexico: State Power and Social Mobilization*. Ithaca, N.Y.: Cornell University Press.

Fox, J. 1996. How Does Civil Society Thicken? The Political Construction of Social Capital in Rural Mexico. *World Development* 24(6): 1089–1103.

Fox, J. 2007. *Accountability Politics: Power and Voice in Rural Mexico*. Oxford: Oxford University Press.

Fox, J, and L. Haight, eds. 2010. *Subsidios para la desigualdad: Las políticas públicas del maíz en México a partir del libre comercio*. Washington, D.C.: Woodrow Wilson International Center for Scholars.

FSC Certificate Database. 2020. FSC. https://us.fsc.org/en-us/market/find-products/fsc-certificate-database.

Fuente Carrasco, M. E., and M. F. Ramos Morales. 2013. "El ecoturismo comunitario en la Sierra Juárez-Oaxaca, México: Entre el patrimonio y la mercancía." *Otra Economía* 7(12): 66–79.

Fukuyama, F. 1996. *Trust: The Social Virtues and the Creation of Prosperity*. New York: Free Press.

Fukuyama, F. 2011. *The Origins of Political Order: From Prehuman Times to the French Revolution*. New York: Farrar, Straus and Giroux.

Fukuyama, F. 2014. *Political Order and Political Decay: From the Industrial Revolution to the Globalization of Democracy*. New York: Farrar, Strauss and Giroux.

Galicia, L., C. Potvin, and Messier. 2015. "Maintaining the High Diversity of Pine and Oak Species in Mexican Temperate Forests: A New Management Approach Combining Functional Zoning and Ecosystem Adaptability." *Canadian Journal of Forest Research* 45(10): 1358–68.

Galletti, H. A. 1998. "The Maya Forest of Quintana Roo: Thirteen Years of Conservation and Community Development." In *Timber, Tourists, and Temples: Conservation and Development in the Maya Forest of Belize, Guatemala, and Mexico,* edited by R. B. Primack, D. B. Bray, H. A. Galletti, and I. Ponciano, 33–46. Washington, D.C: Island Press.

García Aguirre, A. 2014. "Análisis del marco regulatorio en torno a los aprovechamientos forestales maderables y propuesta de mejora." Mexico City: Consejo Civil Mexicano para la Silvicultura Sustentable. http://www.ccmss.org.mx/wp-content/uploads/2014/10/ANALISIS_DEL_MARCO_REGULATORIO_PMF.pdf.

García García, A., C. Marqués Rosano, and L. Merino. 2018. *Bosques con todos: Propuesta de una nueva política forestal.* Ambio/RITA/CCMSS and others. http://www.ccmss.org.mx/acervo/bosques-con-todos-propuesta-de-una-nueva-politica-forestal/.

García-López, G. A. 2013. "Scaling up from the Grassroots and the Top Down: The Impacts of Multi-level Governance on Community Forestry in Durango, Mexico." *International Journal of the Commons* 7(2): 406–31.

García-López, G. A. 2018. "Rethinking Elite Persistence in Neoliberalism: Foresters and Techno-bureaucratic Logics in Mexico's Community Forestry." *World Development* 120 (August): 169–81.

García-López, G. A., and C. Antinori. 2018. "Between Grassroots Collective Action and State Mandates: The Hybridity of Multi-level Forest Associations in Mexico." *Conservation and Society* 16(2): 193–204.

García-Montiel, E., F. Cubbage, A. Rojo-Alboreca, C. Lujan-Álvarez, E. Montiel-Antuna, and J. J. Corral-Rivas. 2017. "An Analysis of Non-state and State Approaches for Forest Certification in Mexico." *Forests* 8(8), 290, 1–15. https://doi.org/10.3390/f8080290.

Garibay Orozco, C. 2008. *Comunalismos y liberalismos campesinos: Identidad comunitaria, empresa social forestal y poder corporado en el México contemporáneo.* Zamora, Michoacán: El Colegio de Michoacán.

Garibay Orozco, C. n.d. *El caso de una pequeña aldea campesina mexicana dueña de una gran industria forestal.* Washington, D.C.: Rights and Resources Initiative. Accessed October 9, 2017. http://rightsandresources.org/wp-content/exported-pdf/elbaconestudiospanish.pdf.

Garzcón Mercado, J. 1975. "Política forestal y el bienestar campesino." *México y sus bosques* 14 (1): 5–24.

Gautam, A. P., G. P. Shavakoti, and E. L. Webb. 2004. "A Review of Forest Policies, Institutions, and Changes in the Resource Condition in Nepal." *International Forestry Review* 6(2):136–48.

George, A. L., and A. Bennett. 2005. *Case Studies and Theory Development in the Social Sciences.* Cambridge, Mass.: MIT Press.

Geréz Fernández, P. 1993. "Marginación social, deforestación y desarrollo rural: Un estudio de caso en el Cofre de Perote, Veracruz, México." Paper presented at the Conference On Common Ground: Interdisciplinary Approaches to Biodiversity Conservation and Land Use Dynamics in the New World, Belo Horizonte, Brazil, December 1–4.

Geréz Fernández, P., and E. Alatorre-Guzmán. 2007. "Los retos de la certificación forestal en la silvicultura comunitaria de México." In *Los bosques comunitarios de Mexico*, edited by D. Bray, L. Merino, and D. Barry, 71–90. Mexico City: Instituto Nacional de Ecología (INE-SEMARNAT).

Gibson, C. C., M. McKean, and E. Ostrom, eds. 2000. *People and Forests: Communities, Institutions, and Governance*. Cambridge, Mass.: MIT Press.

Gibson, C. C., J. T. Williams, and E. Ostrom. 2005. "Local Enforcement and Better Forests." *World Development* 33(2): 273–84.

Gibson, L., T. M. Lee, L. P. Koh, B. W. Brook, T. A. Gardner, J. Barlow . . . and N. S. Sodhi. 2011. "Primary Forests Are Irreplaceable for Sustaining Tropical Biodiversity." *Nature* 478 (7369): 378.

Gijsbers, Wm. n.d. "Manejo forestal en comunidades rurales de Oaxaca: Ahora sí hay una idea de sustentabilidad." P. 55. Manuscript.

Gingrich, Randall W. 1993. "The Political Ecology of Deforestation in the Sierra Madre Occidental of Chihuahua." MA thesis, University of Arizona, Tucson.

Gobierno de México. n.d. "Anuarios estadísticas forestales." Accessed June 10, 2019. https://www.gob.mx/semarnat/documentos/anuarios-estadisticos-forestales.

Gobierno de la República. 2014. *Programa Nacional Forestal 2014–2018*. Mexico City: Gobierno de la República. Accessed May 22, 2018. http://www.conafor.gob.mx:8080/documentos/ver.aspx?articulo=5382&grupo=4.

Goldstein, B. E. 2012. *Collaborative Resilience: Moving Through Crisis to Opportunity*. Cambridge, Mass.: MIT Press.

Gómez-Mendoza, L. and L. Arriaga. 2007. "Modeling the Effect of Climate Change on the Distribution of Oak and Pine Species of Mexico." *Conservation Biology* 21(6): 1545–55.

Gómez-Mendoza, L., E. Vega-Peña, M. Ramírez, J. Palacio-Prieto, and L. Galicia. 2006. "Projecting Land-Use Change Processes in the Sierra Norte of Oaxaca, Mexico." *Applied Geography* 26(3–4, October): 276–90.

González Martínez, Alfonso. 1992. "Los bosques de las tierras mexicanas: La gran tendencia." *El Cotidiano* 8(48): 3–6.

González, M. A., and M. E. Miranda. 2003. "El sistema comunitario para el manejo y protección de la biodiversidad: Cuance Huatulco-Copalita, Oaxaca, México." *Leisa: Revista de Agroecología* 19 (3): 7–9.

Goodwin, N. R. 2003. "Five Kinds of Capital: Useful Concepts for Sustainable Development." February. Global Development and Environment Institute, Tufts University. Working Paper 03–07. https://www.researchgate.net/publication/4861087_03-07_Five_Kinds_of_Capital_Useful_Concepts_for_Sustainable_Development.

Griffiths, D. T. 1958. "Informe sobre silvicultura." In *Aprovechamiento de los recursos forestales: Informes presentados por la misión forestal de la Organización de las Naciones Unidas para la Agricultura y la Alimentación*, 85–111. Mexico City: Banco de México, S.A.

Grindle, M. S. 1977. *Bureaucrats, Politicians, and Peasants in Mexico: A Case Study in Public Policy*. Berkeley: University of California Press.

Griscom, B. W., J. Adams, P. W. Ellis, R. A. Houghton, G. Lomax, D. A. Miteva . . . and P. Woodbury. 2017. "Natural Climate Solutions." *Proceedings of the National Academy of Sciences* 114(44): 11645–50.

Griscom, B. W., and R. Cortez. 2013. "The Case for Improved Forest Management (IFM) as a Priority REDD+ Strategy in the Tropics." http://tropicalconservationscience.mongabay.com/content/v6/TCS-2013_Vol_6%283%29_409-425-Griscom-Cortez.pdf.

Guerra Lizárraga, J. M. 1991. *La explotación forestal en la región de El Salto, Durango*. Thesis for agronomist engineering, Universidad Autónoma Chapingo.

Guerrero, María Teresa, Cyrus Reed, and Brandon Vegter. 2000. *La industria forestal y los recursos naturales en La Sierra Madre de Chihuahua: Impactos sociales, económicos y ecológicos*. Mexico City: Comisión de Solidaridad y Defensa de los Derechos Humanos, A.C., and Texas Center for Policy Studies.

Hahn, T., and B. Nykvist. 2017. "Are Adaptations Self-Organized, Autonomous, and Harmonious? Assessing the Social-Ecological Resilience Literature." *Ecology and Society* 22(1): 12. https://doi.org/10.5751/ES-09026-220112.

Hajjar, R. F., R. A. Kozak, and J. L. Innes. 2012. "Is Decentralization Leading to 'Real' Decision-Making Power for Forest-Dependent Communities? Case Studies from Mexico and Brazil." *Ecology and Society* 17(1): 12. https://doi.org/10.5751/ES-04570-170112.

Halhead, V. 1984. "The Forests of Mexico: The Resource and the Politics of Utilization." Master of philosophy, University of Edinburgh.

Hamilton, N. 1982. *The Limits of State Autonomy: Post-revolutionary Mexico*. Princeton, N.J.: Princeton University Press.

Hardin, G. 1968. "The Tragedy of the Commons." *Science* 162(3859): 1243–48.

Henson, E. 2019. *Agrarian Revolt in the Sierra of Chihuahua, 1959–1965*. Tucson: University of Arizona Press.

Hernández, T., R. Fortín, and R. Butterfield. 2010. "The Impacts of Technical Assistance on a Community Forest Enterprise: The Case of San Bernardino de Milpillas Chico, Mexico." Rainforest Alliance. https://www.profor.info/sites/profor.info/files/Mexico_English_Final_0.pdf.

Hernández-Aguilar, J. A., H. S. Cortina-Villar, L. E. García-Barrios, and M. Á. Castillo-Santiago. 2017. "Factors Limiting Formation of Community Forestry Enterprises in the Southern Mixteca Region of Oaxaca, Mexico." *Environmental Management* 59(3): 490–504.

Hernández Astorga, M. 2015. *El Salto: Crónicas de un pueblo maderero*. Durango: Instituto de Cultura del Estado de Durango.

Hernández-Díaz, J. C., J. C. Rivas, A. Quiñones-Chávez, J. R. Bacon-Sobbe, and B. Vargas-Larreta. 2008. "Evaluación del manejo forestal regular e irregular en bosques de la Sierra Madre Occidental." *Madera y Bosques* 14(3): 25–41.

Hernández-Rodríguez, E., L. Escalera-Vázquez, J. M. Calderón-Patrón, and E. Mendoza. 2019. "Mamíferos medianos y grandes en sitios de tala de impacto reducido y de conservación en la sierra Juárez, Oaxaca." *Revista Mexicana de Biodiversidad* 90(1): 1–10.

Hinojosa Flores, I. D., M. Skutsch, and I. Mustalahti. 2016. "Impacts of Finnish Cooperation in the Mexican Policy Making Process: From the Community Forest Management to the Liberalization of Forest Services." *Forest Policy and Economics* 73:229–38.

Hinojosa Ortiz, Manuel. 1958. *Los bosques de México: Relato de un despilfarro y una injusticia.* Mexico City: Instituto Mexicano de Investigaciones Económicas.

Hodgdon, B. D. 2014. *Market and Finance Options for Ejido Forestry in the Yucatán Peninsula.* Report to the Alianza Mexico REDD+. http://www.monitoreoforestal.gob.mx/repositoriodigital/files/original/9b4e5f50b0e83940decef9227f138f33.pdf.

Hodgdon, B. D., and O. Estrada Murrieta. 2015. *Towards Integrated Community Forest Enterprise: A Case Study of Ejido El Largo y Anexos (Chihuahua, Mexico).* Community Forestry Case Studies No. 3/10. Rainforest Alliance. https://www.rainforest-alliance.org/sites/default/files/2016-08/el-largo-case-study.pdf.

Hornborg, A. 2007. "Introduction: Environmental History as Political Ecology." In *Rethinking Environmental History: World-System History and Global Environmental Change,* edited by A. Hornborg, J. R. McNeill, and J. Martínez-Alier, 1–24. Lanham, Md.: AltaMira Press.

Hostettler, U. 1996. "Milpa Agriculture and Economic Diversification: Socioeconomic Change in a Maya Peasant Society of Central Quintana Roo, 1900–1990s." PhD diss., University of Berne.

Humphries, S. 2010. "Community-Based Forest Enterprises in Brazil and Mexico: Timber Production and Commercialization Models, Market Engagement, and Financial Viability." PhD diss., University of Florida.

Ibarra, J. T., A. Barreau, C. D. Campo, C. I. Camacho, G. J. Martin, and S. R. McCandless. 2011. "When Formal and Market-Based Conservation Mechanisms Disrupt Food Sovereignty: Impacts of Community Conservation and Payments for Environmental Services on an Indigenous Community of Oaxaca, Mexico." *International Forestry Review* 13(3): 318–37.

Ibarra Mendivel, Jorge Luis. 1996. "Cambios recientes en la Constitución mexicana y su impacto sobre La Reforma Agraria." In *Reformando La Reforma Agraria Mexicana,* edited by Laura Randall, 47–63. Mexico City: UAM Unidad Xochimiloc.

Illoldi-Rangel, P., T. Fuller, M. Linaje, C. Pappas, V. Sánchez Cordero, and S. Sarkar. 2008. "Solving the Maximum Representation Problem to Prioritize Areas for the Conservation of Terrestrial Mammals at Risk in Oaxaca." *Diversity and Distributions* 14(3): 493–508.

INEGI. 2009. *Estadísticas históricas de México.* 2nd ed. Aguascalientes, Mexico: Instituto Nacional de Estadísticas, Geografía e Informática.

INGEI. 1988–90, 1992, 2002–16. *Anuario Estadística y Geográfica de los Estados Unidos Mexicanos*. Aguascalientes, Mexico: Instituto Nacional de Estadística, Geografía e Informática.

Informe de labores del gobierno de Pedro Joaquín Coldwell, 1987–1990. Accessed September 2004 in library Chetumal, QR.

IPCC. 2018. "Summary for Policymakers." In *Global Warming of 1.5°C. An IPCC Special Report on the Impacts of Global Warming of 1.5°C Above Pre-industrial Levels and Related Global Greenhouse Gas Emission Pathways, in the Context of Strengthening the Global Response to the Threat of Climate Change, Sustainable Development, and Efforts to Eradicate Poverty*, edited by V. Masson-Delmotte, P. Zhai, H.-O. Pörtner, D. Roberts, J. Skea, P. R. Shukla, A. Pirani, W. Moufouma-Okia, C. Péan, R. Pidcock, S. Connors, J. B. R. Matthews, Y. Chen, X. Zhou, M. I. Gomis, E. Lonnoy, T. Maycock, M. Tignor, and T. Waterfield, 1–24. Geneva: World Meteorological Organization.

ITS Global. 2011. "The Economic Contribution of Indonesia's Forest-Based Industries." https://static1.squarespace.com/static/562c7435e4b01a45f69f18f9/t/5725d87707eaa04b68665481/1462098045495/The+Economic+Contribution+of+Indonesia%E2%80%99s+Forest-Based+Industries+-+Report+Annex+%282011%29.pdf.

ITTO (International Tropical Timber Association). *Tropical Timber Market Report* 20(5). March 1–15. https://www.lifeforestry.com/fileadmin/template/main/pdf/itto/ITTO-Report-1603-01-15.pdf.

Jones, D. C., and J. Svejnar. 1982. *Participatory and Self-Managed Firms: Evaluation of Economic Performance*. Lexington, Mass.: Lexington Books.

Jönsson, C. 1986. "Interorganization Theory and International Organization." *International Studies Quarterly* 30(1): 39–57.

Jorgensen, M., and J. B. Taylor. 2000. *What Determines Indian Economic Success?: Evidence from Tribal and Individual Indian Enterprises*. Cambridge, Mass.: Harvard University.

Jurjonas, M., and E. Seekamp. 2019. "Balancing Carbon Dioxide: A Case Study of Forest Preservation, Out-migration, and Afforestation in the Pueblos Mancomunados of Oaxaca, Mexico." *Journal of Sustainable Forestry* 38(7): 697–714.

Kaimowitz, D. 2004. "Market Options and Barriers for Timber and Sawnwood from Michoacán, Oaxaca, Guerrero, Campeche and Quintana Roo (English Summary)." Washington, D.C.: Forest Trends. https://www.forest-trends.org/publications/market-options-and-barriers-for-timber-and-sawnwood-from-michoacan-oaxaca-guerrero-campeche-and-quinatana-roo-english-summary/.

Karmann, M., and A. Smith. 2009. "FSC Reflected in Scientific and Professional Literature: Literature Study on the Outcomes and Impacts of FSC Certification." April. Bonn, Germany: FSC International Center. https://www.fsc-deutschland.de/preview.fsc-impact-study-engl.a-1341.pdf.

Kashwan, P. 2017. *Democracy in the Woods: Environmental Conservation and Social Justice in India, Tanzania, and Mexico*. Oxford: Oxford University Press.

Kearney, M. 1986. *The Winds of Ixtepeji: World View and Society in a Zapotec Town*. Prospect Heights, Ill.: Waveland Press.

Kerr, J., M. Vardhan, and R. Jindal. 2012. "Prosocial Behavior and Incentives: Evidence from Field Experiments in Rural Mexico and Tanzania." *Ecological Economics* 73(January): 220–27.

Kim, S., M. J. Karlesky, C. G. Myers, and T. Schifeling. 2016. "Why Companies Are Becoming B Corporations." *Harvard Business Review* 17(1): 6–15.

Kinzig, A. P., P. Ryan, M. Etienne, H. Allison, T. Elmqvist, and B. H. Walker. 2006. "Resilience and Regime Shifts: Assessing Cascading Effects." *Ecology and Society* 11(1): 20. http://www.ecologyandsociety.org/vol11/iss1/art20.

Klemperer, W. D. 1996. *Forest Resource Economics and Finance.* New York: McGraw-Hill.

Klivak, L. 2016. "Que salgamos más adelante: Identity, Community, and the Desire for Development in Oaxaca, Mexico." PhD diss., Syracuse University, Syracuse, N.Y. https://surface.syr.edu/cgi/viewcontent.cgi?article=1518&context=etd.

Klooster, D. J. 1997. "Conflict in the Commons: Commercial Forestry and Conservation in Mexican Indigenous Communities." PhD diss., University of California, Los Angeles.

Klooster, D. J. 2000. "Institutional Choice, Community, and Struggle: A Case Study of Forest Co-Management in Mexico." *World Development* 28(1): 1–20.

Klooster, D. J. 2002. "Toward Adaptive Community Forest Management: Integrating Local Forest Knowledge with Scientific Forestry." *Economic Geography* 78(1): 43–70.

Klooster, D. J. 2006. "Environmental Certification of Forests in Mexico: The Political Ecology of a Nongovernmental Market Intervention." *Annals of the Association of American Geographers* 96(3): 541–65.

Klooster, D. J., and S. Ambinakudige. 2005. "The Global Significance of Mexican Community Forestry." In *The Community Forests of Mexico: Managing for Sustainable Landscapes*, edited by D. B. Bray, L. Merino-Pérez, and D. Barry, 305–34. Austin: University of Texas Press.

Klooster, D. J., and O. Masera. 2000. "Community Forest Management in Mexico: Carbon Mitigation and Biodiversity Conservation Through Rural Development." *Global Environmental Change* 10(4): 259–72.

Klooster, D., and A. Mercado-Celis. 2016. "Sustainable Production Networks: Capturing Value for Labour and Nature in a Furniture Production Network in Oaxaca, Mexico." *Regional Studies* 50(11): 1889–1902.

Klooster, D., R. Taravella, and B. D. Hodgdon. 2015. "Striking the Balance: Adapting Community Forest Enterprise to Meet Market Demands. A Case Study of TIP Muebles (Oaxaca, Mexico)." *Community Forestry Case Studies No. 7.* Rainforest Alliance / FOMIN. http://www.rainforest-alliance.org/case-studies/tip-muebles.

Knowlton, R. J. 1998. "El ejido mexicano en el siglo XIX." *Historia Mexicana* 48(1): 171–96.

Kolb, M., and L. Galicia. 2012. "Challenging the Linear Forestation Narrative in the Neo-tropic: Regional Patterns and Processes of Deforestation and Regeneration in Southern Mexico." *Geographical Journal* 178(2): 147–61.

Konrad, H. W. 1987. "Capitalismo y trabajo en los bosques de las tierras bajas tropicales mexicanas: El caso de la industria del chicle." *Historia Mexicana* 36(3): 465–505.

Konrad, H. W. 1991. "Capitalism on the Tropical-Forest Frontier: Quintana Roo, 1880s to 1930." In *Land, Labor, and Capital in Modern Yucatan: Essays in Regional History and Political Economy*, edited by J. G. Brannon, 67–94. Tuscaloosa: University of Alabama Press.

Krishna, A. 2000. "Creating and Harnessing Social Capital." In *Social Capital: A Multi-faceted Perspective*, edited by P. Dasgupta and I. Serageldin, 71–93. Washington, D.C.: World Bank.

Lammertink, J. M., J. A. Rojas-Tomé, F. M. Casillas-Orona, and R. L. Otto. 1996. "Status and Conservation of Old-Growth Forests and Endemic Birds in the Pine-Oak Zone of the Sierra Madre Occidental, Mexico." *Verslagen en Technische Gegevens* 69(1): 1–89.

Larson, A. M., and F. Soto. 2008. "Decentralization of Natural Resource Governance Regimes." *Annual Review of Environment and Resources* 33:213–39.

Lartigue, F. 1983. *Indios y bosques: Políticas forestales y comunales en La Sierra Tarahumara*. Ediciones de la Casa Chata. 1st ed. Vol. 19. Mexico City: Centro de Investigaciones y Estudios Superiores en Antropología Social.

Laurance, W. F., D. C. Useche, J. Rendeiro, M. Kalka, C. J. Bradshaw, S. P. Sloan . . . and V. Arroyo-Rodriguez. 2012. "Averting Biodiversity Collapse in Tropical Forest Protected Areas." *Nature* 489(7415): 290.

Lemke, J. 2019. "10 Tips to Facilitate Collective Action from Elinor and Vincent Ostrom." Libertarianism.org. https://www.libertarianism.org/columns/10-tips-facilitate-collective-action-from-elinor-vincent-ostrom.

Leslie, H. M., X. Basurto, M. Nenadovic, L. Sievanen, K. C. Cavanaugh, J. Cota-Nieto . . . and S. Nagavarapu. 2015. "Operationalizing the Social-Ecological Systems Framework to Assess Sustainability." *Proceedings of the National Academy of Sciences* 112(19): 5979–84.

Levin, S., T. Xepapadeas, A. S. Crépin, J. Norberg, A. De Zeeuw, C. Folke . . . and P. Ehrlich. 2013. "Social-Ecological Systems as Complex Adaptive Systems: Modeling and Policy Implications." *Environment and Development Economics* 18(2): 111–32.

Liu, J., T. Dietz, S. R. Carpenter, M. Alberti, C. Folke, E. Moran . . . and E. Ostrom. 2007. "Complexity of Coupled Human and Natural Systems." *Science* 317(5844): 1513–16.

Liu, J., V. Hull, M. Batistella, R. DeFries, T. Dietz, F. Fu . . . and L. A. Martinelli. 2013. "Framing Sustainability in a Telecoupled World." *Ecology and Society* 18(2): 26. https://www.ecologyandsociety.org/vol18/iss2/art26/.

López Álvarez, M. G. 1994. "El Largo y sus anexos: Una historia de explotación en Chihuahua." *La Jornada del Campo*, June 28, 11.

López Arzola, R., and P. Geréz Fernández. 1993. "The Permanent Tension." *Cultural Survival Quarterly* 17(1): 42–44.

López Pardo, G., and B. Palomino Villavicencio. 2008. "Políticas públicas y ecoturismo en comunidades indígenas de México." *Teoría y Praxis* 4(5): 33–50.

López Santos, Otoniel. 1948. "La explotación de maderas de especies coníferas en México." Thesis in Forest Engineering. Universidad Nacional Autónoma de México.

Lynch, J. F., and D. F. Whigham. 1995. "The Role of Habitat Disturbance in the Ecology of Overwintering Birds in the Yucatan Peninsula." In *Conservation of Neotropical Migratory Birds in Mexico*, edited by M. H. Wilson, S. Sader, and A. Estrada, 199–214. Miscellaneous publication 727. Orono: Maine Agricultural and Forest Experiment Station.

Macqueen, D., A. Bolin, M. Greijmans, S. Grouwels, and S. Humphries. 2020. "Innovations Towards Prosperity Emerging in Locally Controlled Forest Business Models and Prospects for Scaling Up." *World Development* 125 (July): n.p. https://doi.org/10.1016/j.worlddev.2018.08.004.

Madrid, L., J. M. Núñez, G. Quiroz, and Y. Rodríguez. 2009. "La propiedad social forestal en México." *Investigación Ambiental* 1(2): 179–96.

Madrid, S. 2019. "El reto de la CONAFOR." *La Jornada*. https://www.jornada.com.mx/2019/01/19/cam-conafor.html.

Martin, G. J., C. I. Camacho Benavides, C. A. Del Campo García, S. Anta Fonseca, Chapela Mendoza, and M. A. González Ortiz. 2011. "Indigenous and Community Conserved Areas in Oaxaca, Mexico." *Management of Environmental Quality: An International Journal* 22(2): 250–66.

Martin, G., C. del Campo, C. Camacho, G. Espinoza-Sauceda, and X. Zolueta Juan. 2010. "Negotiating the Web of Law and Policy: Community Designation of Indigenous and Community Conserved Areas in Mexico." *Policy Matters* 17:195–204.

Markiewicz, D. 1993. *The Mexican Revolution and the Limits of Agrarian Reform, 1915–1946*. Boulder: Lynne Rienner Publishers.

Martínez-Alier, J. M. 2002. *The Environmentalism of the Poor: A Study of Ecological Conflicts and Valuation*. Cheltenham, UK: Edward Elgar.

Martinez-Alier, J. 2004. "Ecological Distribution Conflicts and Indicators of Sustainability." *International Journal of Political Economy* 34(1): 13–30.

Masera, O. R., M. J. Ordóñez, and R. Dirzo. 1997. "Carbon Emissions from Mexican Forests: Current Situation and Long-Term Scenarios." *Climatic Change* 35(3): 265–95.

Mathews, A. S. 2006. "Building the Town in the Country: Official Understandings of Fire, Logging and Biodiversity in Oaxaca, Mexico, 1926–2004." *Social Anthropology* 14(3): 335–59.

Mathews, A. S. 2011. *Instituting Nature: Authority, Expertise, and Power in Mexican Forests*. Cambridge, Mass.: MIT Press.

Matson, P., W. C. Clark, and K. Andersson. 2016. *Pursuing Sustainability: A Guide to the Science and Practice*. Princeton, N.J.: Princeton University Press.

McKean, M. A. 1992. "Success on the Commons: A Comparative Examination of Institutions for Common Property Resource Management." *Journal of Theoretical Politics* 4(3): 247–81.

McKean, M. A. 2000. "Common Property: What Is It, What Is It Good for, and What Makes It Work?" In *People and Forests: Communities, Institutions, and Governance*, edited by M. A. McKean, C. C. Gibson, and E. Ostrom, 27–55. Cambridge, Mass.: MIT Press.

McAfee, K. and E. N. Shapiro. 2010. "Payments for Ecosystem Services in Mexico: Nature, Neoliberalism, Social Movements, and the State." *Annals of the Association of American Geographers* 100(3): 579–99.

McGinnis, M., and E. Ostrom. 2014. "Social-Ecological System Framework: Initial Changes and Continuing Challenges." *Ecology and Society* 19(2): 30. https://doi.org/ 10.5751/ES-06387-190230.

Meave, J. A., A. Rincón-Gutiérrez, G. Ibarra-Manríquez, C. Gallardo-Hernández, and M. A. Romero. 2017. "Checklist of the Vascular Flora of a Portion of the Hyper-humid Region of La Chinantla, Northern Oaxaca Range, Mexico." *Botanical Sciences* 95(4): 722–59.

Mendoza, E., and Dirzo, R. 1999. "Deforestation in Lacandonia (southeast Mexico): Evidence for the Declaration of the Northernmost Tropical Hot-Spot." *Biodiversity and Conservation* 8(12): 1621–41.

Mendoza Medina, R. 1976. "La política forestal en el sector ejidal y comunal." *Revista del México Agrario* 9(2): 35–69.

Merino Pérez, L. 1997. "Organización Social de la Producción Forestal Comunitaria: Siete estudios de caso." In *Semillas para el cambio en el campo: Medio ambiente, mercados y organización campesina*, edited by L. Paré, D. B. Bray, J. Burstein, and S. Martínez Vásquez, 141–54. Mexico City: UNAM; IIS; La Sociedad de Solidaridad Social "Sansekan Tinemi" y Saldebas; Servicios de Apoyo Local al Desarrollo de Base en México.

Merino Pérez, L., G. Alatorre, B. Cabarle, F. Chapela, and S. Madrid. 1997. *El manejo forestal comunitario en México y sus perspectivas de sustentabilidad*. 1st ed. Cuernavaca, Morelos: Centro Regional de Investigaciones Multidisciplinarias, UNAM.

Merino Pérez, L. 2004. *Conservación o deterioro: El impacto de las políticas públicas en las instituciones comunitarias y en los usos de los bosques en México*. Mexico City: Instituto Nacional de Ecología / Consejo Civil Mexicano para la Silvicultura Sostenible.

Merino Pérez, L., and M. Hernández Apolinar. 2004. "Destrucción de instituciones comunitarias y deterioro de los bosques en la Reserva de la Biósfera Mariposa Monarca, Michoacán, México." *Revista Mexicana de Sociología* 66(2): 261–309.

Merino Pérez, L., G. Ortiz, and J. Rodríguez. 2013. "La política forestal." In *Encuentros y desencuentros: Las comunidades forestales y las políticas públicas en tiempos de transición*, edited by L. Merino Pérez and G. Ortiz Merino, 101–48. UNAM / Instituto de Investigaciones Sociales. Miguel Angel Porrúa: Mexico City.

Merino Pérez, L., and G. Segura. 2005. "Forest and Conservation Policies and Their Impact on Forest Communities in Mexico." In *The Community Forests of Mexico: Managing for Sustainable Landscapes*, edited by D. B. Bray, L. Merino-Pérez, and D. Barry, 49–70. Austin: University of Texas Press.

Merlet, M. 2015. "Community Forests: Document for Discussion Between Fern and Its Partners." https://www.agter.org/bdf/_docs/merlet_fern_2017_community_forests _discussion_document_final.pdf.

Messier, C., K. Puettmann, R. Chazdon, K. P. Andersson, V. A. Angers, L. Brotons, . . . and Levin, S. A. 2015. "From Management to Stewardship: Viewing Forests as Complex Adaptive Systems in an Uncertain World." *Conservation Letters* 8(5): 368–77.

Millar, C. I., N. L. Stephenson, and S. L. Stephens. 2007. "Climate Change and Forests of the Future: Managing in the Face of Uncertainty." *Ecological Applications* 17(8): 2145–51.

Miteva, D. A., P. W. Ellis, E. A. Ellis, and B. W. Griscom. 2019. "The Role of Property Rights in Shaping the Effectiveness of Protected Areas and Resisting Forest Loss in the Yucatan Peninsula." *PloS One* 14(5): e0215820.

Mittermeier, R., P. Robles Gil, and C. G. Mittermeier, eds. 1997. *Megadiversity: Earth's Biologically Wealthiest Nations*. Mexico City: Cemex.

Moguel, J. 1989. "La cuestión agraria en el período 1950–1970." In *Política estatal y conflictos agrarios 1950–1970*. Vol. 8, *Historia de la cuestión agraria mexicana*, edited by J. Moguel, 112–34. Mexico City: Siglo Veintiuno Editores / Centro de Estudios Históricos del Agrarismo en México.

Moguel, J., and P. López Sierra. 1990. "Política agraria y modernización capitalista." In *Historia de la cuestión agraria: Los tiempos de la crisis (segunda parte), 1970–1982*, edited by J. Moguel, 321–76. Mexico City: Siglo Veintiuno editores / CEHAM.

Molnar, A., D. Gomes, R. Sousa, N. Vidal, R. F. Hojer, L. A. Arguelles . . . and A. White. 2008. "Community Forest Enterprise Markets in Mexico and Brazil: New Opportunities and Challenges for Legal Access to the Forest." *Journal of Sustainable Forestry* 27(1–2): 87–121.

Monterroso, I., and D. Barry. 2012. "Legitimacy of Forest Rights: The Underpinnings of the Forest Tenure Reform in the Protected Areas of Petén, Guatemala." *Conservation and Society* 10(2): 136–50.

Moran, Emilio F., and Elinor Ostrom, eds. 2005. *Seeing the Forest and the Trees: Human-Environment Interactions in Forest Ecosystems*. Cambridge, Mass.: MIT Press.

Morrow, C. E., and R. W. Hull. 1996. "Donor-Initiated Common Pool Resource Institutions: The Case of the Yanesha Forestry Cooperative." *World Development* 24(10): 1641–57.

Mota-Villanueva, J. L. 2002. "Estudio de caso de integración horizontal: Sociedad de Productores Forestales Ejidales de Quintana Roo, S.C. Banco Interamericano de Desarrollo." http://www.ccmss.org.mx/acervo/integracion-horizontal-sociedad-de-productores-forestales-ejidales-de-quintana-roo/.

Muñoz-Piña, C., A. De Janvry, and E. Sadoulet. 2003. "Recrafting Rights over Common Property Resources in Mexico." *Economic Development and Cultural Change* 52(1): 129–58.

Muñoz-Piña, C., A. Guevara, J. M. Torres, and J. Braña. 2008. "Paying for the Hydrological Services of Mexico's Forests: Analysis, Negotiations and Results." *Ecological Economics* 65(4): 725–36.

Muradian, R., E. Corbera, U. Pascual, N. Kosoy, and P. H. May. 2010. "Reconciling Theory and Practice: An Alternative Conceptual Framework for Understanding Payments for Environmental Services." *Ecological Economics* 69(6): 1202–8.

Nagendra, H. 2007. "Drivers of Reforestation in Human-Dominated Forests." *Proceedings of the National Academy of Sciences* 104(39): 15218–223.

Navarro-Martínez, A., E. A. Ellis, I. Hernández-Gómez, J. A. Romero-Montero, and O. Sánchez-Sánchez. 2018. "Distribution and Abundance of Big-Leaf Mahogany (*Swietenia macrophylla*) on the Yucatan Peninsula, Mexico." *Tropical Conservation Science* 11. https://doi.org/10.1177/1940082918766875.

Navia-Antezana, J., M. C. Marín-Togo, and I. Cumana-Navia. 2018. "El manejo forestal comunitario en Michoacán." In *Las empresas sociales forestales en México: Claroscuros y aprendizajes*, edited by G. Chapela, 77–98. Mexico City: Consejo Civil Mexicano para la Silvicultura Sostenible, AC.

Nepstad, D. C., C. M. Stickler, B. D. Filho, and F. Merry. 2008. "Interactions Among Amazon Land Use, Forests and Climate: Prospects for a Near-Term Forest Tipping Point." *Philosophical Transactions of the Royal Society B: Biological Sciences* 363(1498): 1737–46.

Nieratka, L. R., D. B. Bray, and P. Mozumder. 2015. "Can Payments for Environmental Services Strengthen Social Capital, Encourage Distributional Equity, and Reduce Poverty?" *Conservation and Society* 13, no. 4 (October–December): 345–55.

North, D. C. 1993. "Economic Performance Through Time." Nobel Prize Lecture. https://www.nobelprize.org/prizes/economic-sciences/1993/north/lecture/.

Núñez Rodríguez, V. R. n.d. "A cien años del inicio de La Reforma Agraria en México." REDPOL No. 9. Departamento de Administración: Universidad Autónoma Metropolitana–Azcapotzalco. http://redpol.azc.uam.mx/index.php/2016-05-03-05 -08-25/redpol-9.

Ocampo Arista, S. 2017. "Habitantes de Coyuca, exiliados y a merced del crimen organizado." *La Jornada*, November 3, 24. https://www.jornada.com.mx/2017/11/03/estados/024n1est.

Ochoa-Gaona, S., and González-Espinosa, M. 2000. "Land Use and Deforestation in the Highlands of Chiapas, Mexico." *Applied Geography* 20(1): 17–42.

Ochoa-Ochoa, L., J. N. Urbina-Cardona, L. B. Vazquez, O. Flores-Villela, and J. Bezaury-Creel. 2009. "The Effects of Governmental Protected Areas and Social Initiatives for Land Protection on the Conservation of Mexican Amphibians." *PLoS One* 4(9): e6878. http://dx.doi.org/10.1371/journal.pone.0006878.

Ojha, H. R., R. Ford, R. J. Keenan, D. Race, D. C. Vega, H. Baral, and P. Sapkota. 2016. "Delocalizing Communities: Changing Forms of Community Engagement in Natural Resources Governance." *World Development* 87:274–90.

Oliver, C. D., and B. C. Larson. 1996. *Forest Stand Dynamics: Updated Edition.* New York: John Wiley and Sons.

Olsson, P., C. Folke, and F. Berkes. 2004. Adaptive Comanagement for Building Resilience in Social-Ecological Systems. *Environmental Management* 34(1): 75–90.

Orta-Martínez, M., and M. Finer. 2010. "Oil Frontiers and Indigenous Resistance in the Peruvian Amazon." *Ecological Economics* 70(2): 207–18.

Osborne, T., and E. Shapiro-Garza. 2018. "Embedding Carbon Markets: Complicating Commodification of Ecosystem Services in Mexico's Forests." *Annals of the American Association of Geographers* 108(1): 88–105.

Ostrom, E. 1990. *Governing the Commons: The Evolution of Institutions for Collective Action.* New York: Oxford University Press.

Ostrom, E. 2000. "Collective Action and the Evolution of Social Norms." *Journal of Economic Perspectives* 14(3): 137–58.

Ostrom, E. 2005. *Understanding Institutional Diversity*. Princeton, N.J.: Princeton University Press.

Ostrom, E. 2007. "A Diagnostic Approach for Going Beyond Panaceas." *Proceedings of the National Academy of Sciences* 104 (39): 15181–187. https://doi.org/10.1073/pnas.0702288104.

Ostrom, E. 2009. "A General Framework for Analyzing Sustainability of Social-Ecological Systems." *Science* 325(5939): 419–22.

Ostrom, E. 2010. "Polycentric Systems for Coping with Collective Action and Global Environmental Change." *Global Environmental Change* 20(4): 550–57.

Ostrom, E. 2011. "Background on the Institutional Analysis and Development Framework." *Policy Studies Journal* 39(1): 7–27.

Ostrom, E., and T. K. Ahn. 2003. "A Social Science Perspective on Social Capital: Social Capital and Collective Action." *Revista Mexicana de Sociología* 65(1): 155–233.

Ostrom, E., and M. A. Janssen. 2004. "Multi-level Governance and Resilience of Social-Ecological Systems." In *Globalisation, Poverty and Conflict*, edited by M. Spoor, 239–59. Dordrecht, Netherlands: Springer.

Ostrom, E., and H. Nagendra. 2006. "Insights on Linking Forests, Trees, and People from the Air, on the Ground, and in the Laboratory." *Proceedings of the National Academy of Sciences* 103(51): 19224–231.

Ovando H., Francisco Xavier. 1979. "Los diversos sistemas de organización en la rama forestal." *México y Sus Bosques* 18(2): 22–28.

Pacheco-Aquino, G., E. Durán Medina, and J. A. B. Ordóñez-Díaz. 2015. "Estimación del carbono arbóreo en el área de manejo forestal de Ixtlán de Juárez, Oaxaca, México." *Revista Mexicana de Ciencias Forestales* 6(29): 126–45.

Palomino Villavicencio, C. B., and A. G. López Pardo. 2011. "Ecoturismo indígena en Quintana Roo, México: Estudio de caso kantemo." *Tourism and Management Studies* (1): 990–98.

Paudel, N. S., I. Monterroso, and P. Cronkleton. 2010. "Community Networks, Collective Action and Forest Management Benefits." In *Forests for People: Community Rights and Forest Tenure Reform*, edited by A. M. Larson, D. Barry, G. R. Dahal, and C. J. P. Colfer, 116–36. Sterling, Va.: Earthscan.

Pazos-Almada, B., and D. B. Bray. 2018. "Community-Based Land Sparing: Territorial Land-Use Zoning and Forest Management in the Sierra Norte of Oaxaca, Mexico." *Land Use Policy* 78 (November): 219–26.

Pérez-Cirera, V. 2004. "Power, Heterogeneity and Local Common Property Resources Governance: An Economic and Political Analysis of Community Forestry in La Sierra Tarahumara, Mexico." PhD diss., University of York (UK), Departments of Environment and Politics.

Pérez-Cirera, V., and J. C. Lovett. 2006. "Power Distribution, the External Environment and Common Property Forest Governance: A Local User Groups Model." *Ecological Economics* 59(3): 341–52.

Perez-Verdin, G., J. Monarrez-Gonzalez, A. Tecle, and M. Pompa-Garcia. 2018. "Evaluating the Multi-functionality of Forest Ecosystems in Northern Mexico." *Forests* 9(4): 178.

Persha, L., A. Agrawal, and A. Chhatre. 2011. "Social and Ecological Synergy: Local Rulemaking, Forest Livelihoods, and Biodiversity Conservation." *Science* 331(6024): 1606–8.

Porritt, J. 2007. *Capitalism: As If the World Mattered.* 2nd ed. London: Earthscan.

Porter-Bolland, Luciana, Edward A. Ellis, Manuel R. Guariguata, Isabel Ruiz-Mallén, Simoneta Negrete-Yankelevich, and Victoria Reyes-García. 2012. "Community Managed Forests and Forest Protected Areas: An Assessment of Their Conservation Effectiveness Across the Tropics." *Forest Ecology and Management* 268 (2012): 6–17.

Pretty, J. 2003. "Social Capital and the Collective Management of Resources." *Science* 302(5652): 1912–14.

"Primera planta cogeneración con biomasa en México." 2020. Expo Biomasa. https://www.expobiomasa.com/content/primera-planta-cogeneracion-con-biomasa-en-mexico.

PROCYMAF. 2000. *Proyecto de Conservación y Manejo Sustentable de Recursos Forestales en México: Balance de tres años de ejecución.* Mexico City: SEMARNAP.

Putz, F. E., P. Sist, T. Fredericksen, and D. Dykstra. 2008. "Reduced-Impact Logging: Challenges and Opportunities." *Forest Ecology and Management* 256(7): 1427–33.

Putz, F. E., T. Baker, B. W. Griscom, T. Gopalakrishna, A. Roopsind, P. M. Umunay . . . and P. W. Ellis. 2019. "Intact Forest in Selective Logging Landscapes in the Tropics." *Frontiers in Forests and Global Change* 2:30. doi: 10.3389/ffgc.2019.00030.

Racelis, A. E., and J. A. Barsimantov. 2008. "The Management of Small Diameter, Lesser-Known Hardwood Species as Polewood in Forest Communities of Central Quintana Roo, Mexico." *Journal of Sustainable Forestry* 27(1–2): 122–44.

Radachowsky, J., V. H. Ramos, R. McNab, E. H. Baur, and N. Kazakov. 2012. "Forest Concessions in the Maya Biosphere Reserve, Guatemala: A Decade Later." *Forest Ecology and Management* 268 (March 15): 18–28.

Raj Sharma, A. 2009. "Impact of Community Forestry on Income Distribution in Nepal." PhD diss., Tribhuvan University, Faculty of Humanities and Social Sciences.

Ramirez-Reyes, C., K. R. Sims, P. Potapov, and V. C. Radeloff. 2018. "Payments for Ecosystem Services in Mexico Reduce Forest Fragmentation." *Ecological Applications* 28(8): 1982–97.

Randolph, J. C., G. M. Green, J. Belmont, T. Burcsu, and D. Welch. 2005. "Forest Ecosystems and the Human Dimension." In *Seeing the Forest and the Trees: Human-Environment Interactions in Forest Ecosystems,* edited by Emilio F. Moran and Elinor Ostrom, 105–26. Cambridge, Mass.: MIT Press.

Rands, M. R., W. M. Adams, L. Bennun, S. H. Butchart, A. Clements, D. Coomes . . . and W. J. Sutherland. 2010. "Biodiversity Conservation: Challenges Beyond 2010." *Science* 329(5997): 1298–1303.

Rasolofoson, R. A., P. J. Ferraro, C. N. Jenkins, and J. P. Jones. 2015. "Effectiveness of Community Forest Management at Reducing Deforestation in Madagascar." *Biological Conservation* 184 (April): 271–77.

Redclift, M. 2004. *Chewing Gum: The Fortunes of Taste.* Routledge: New York.

Rehfeldt, G. E., N. L. Crookston, C. Sáenz-Romero, and E. M. Campbell. 2012. "North American Vegetation Model for Land-Use Planning in a Changing Climate: A Solution to Large Classification Problems." *Ecological Applications* 22(1): 119–41.

Rice, R. E., C. A. Sugal, S. M. Ratay, and G. A. Fonseca. 2001. "Sustainable Forest Management: A Review of Conventional Wisdom." *Advances in Applied Biodiversity Science*, no. 3:1–29. Washington, D.C.: CABS / Conservation International.

Rietkerk, M., S. C. Dekker, P. C. De Ruiter, and J. van de Koppel. 2004. "Self-Organized Patchiness and Catastrophic Shifts in Ecosystems." *Science* 305(5692): 1926–29.

Rights and Resources Initiative. 2018. "At a Crossroads: Consequential Trends in Recognition of Community-Based Forest Tenure From 2002–2017." Washington, D.C.: Rights and Resources Initiative. https://rightsandresources.org/en/publication/at-a -crossroads-trends-in-recognition-of-community-based-forest-tenure-from-2002 -2017/#.W6vBqCBRc2w.

Robles Berlanga, H. M. 2012. "Ejidos y comunidades en México: Problemas y perspectivas." *Memoria del Seminario: Propiedad Social y Servicios Ambientales*. November 8, 2011. https:// www.researchgate.net/publication/263275643_Memorias_del_Seminario_Propiedad _Social_y_Servicios_Ambientales_8_de_noviembre_de_2011_Ciudad_de_Mexico.

Robson, J. P. 2007. "Local Approaches to Biodiversity Conservation: Lessons from Oaxaca, Southern Mexico." *International Journal of Sustainable Development* 10(3): 267–86.

Robson, J. P. 2019. "Adaptive Governance or Cultural Transformation? The Monetization of Uses y Costumbres in Santiago Comaltepec." In *Communities Surviving Migration: Village Governance, Environment and Cultural Survival in Indigenous Mexico*, edited by J. P. Robson, D. Klooster, and J. Hernández-Díaz, n.p. Routledge: New York.

Robson, J. P., and F. Berkes. 2011. "Exploring Some of the Myths of Land Use Change: Can Rural to Urban Migration Drive Declines in Biodiversity?" *Global Environmental Change* 21(3): 844–54.

Robson, J. P., and D. J. Klooster. 2018. "Migration and a New Landscape of Forest Use and Conservation." *Environmental Conservation* 46 (March): 1–8.

Rodríguez Salazar, J, J. A. Anguiano Martínez, and D. B. Bray. 2015. "La creación del PROCYMAF como estrategia de apoyo al desarrollo forestal comunitario en México." In *Desarrollo Forestal Comunitario*, edited by J. M. Torres Rojo, 119–44. Mexico City: Centro de Investigación y Docencia Económica.

Rojas-Soto, O. R., V. Sosa, and J. F. Ornelas. 2012. "Forecasting Cloud Forest in Eastern and Southern Mexico: Conservation Insights Under Future Climate Change Scenarios." *Biodiversity and Conservation* 21(10): 2671–90.

Rosas-Baños, M., and R. Lara-Rodríguez. 2013. "Desarrollo endógeno local sustentable y propiedad común: San Pedro El Alto, México." *Cuadernos de Desarrollo Rural* 10(71): 59–80.

Roy Chowdhury R., and L. Schneider. 2004. "Land Cover and Land Use: Classification and Change Analysis." In *Integrated Land-Change Science and Tropical Deforestation in the Southern Yucatán*, edited by Turner BL II, J. Geoghegan, D. R. Foster, n.p. New York: Oxford University Press.

Ruiz-Talonia, L. F., N. M. Sánchez-Vargas, J. S. Bayuelo-Jiménez, S. I. Lara-Cabrera, and C. Sáenz-Romero. 2014. "Altitudinal Genetic Variation Among Native Pinus Patula Provenances: Performance in Two Locations, Seed Zone Delineation and Adaptation to Climate Change." *Silvae Genetica* 63(1–6): 139–48.

Rzedowski, J. 1988. *Vegetación de México*. 4th ed. Mexico City: Limusa.

Rzedowski, J. 1990. "Vegetación potencial." Conabio. Accessed March 14, 2020. http://www.conabio.gob.mx/informacion/gis/layouts/vpr4mgw.png.

Sáenz-Romero, C., G. E. Rehfeldt, N. L. Crookston, P. Duval, R. St-Amant, J. Beaulieu, and B. A. Richardson. 2010. "Spline Models of Contemporary, 2030, 2060 and 2090 Climates for Mexico and Their Use in Understanding Climate-Change Impacts on the Vegetation." *Climatic Change* 102(3–4): 595–623.

Sahoo, U. K., and G. Sahu. 2019. "Twelve Years Later: Implementation of the Scheduled Tribes and Other Traditional Forest Dwellers (Recognition of Forest Rights) Act 2006." *Indian Journal of Social Work* 80(4): 423–38.

Sánchez Jacinto, L. V. 2011. "El OTC y la construcción de arreglos institucionales a nivel de cuenca: El caso del SICOBI." MS PowerPoint presentation. GAIA, A.C. https://docplayer.es/79576246-El-otc-y-la-construccion-de-arreglos-institucionales-a-nivel-de-cuenca-el-caso-del-sicobi.html.

Sanderson, S. E. 1981. *Agrarian Populism and the Mexican State*. Berkeley: University of California Press.

Sarukhán, J., and R. Dirzo. 2001. "Biodiversity-Rich Countries." In *Encyclopedia of Biodiversity*, edited by S. A. Levin. Vol. 1:419–36. San Diego: Academic Press.

Sarukhán, J., T. Urquiza-Haas, P. Koleff, J. Carabias, R. Dirzo, E. Ezcurra . . . and J. Soberón. 2014. "Strategic Actions to Value, Conserve, and Restore the Natural Capital of Megadiversity Countries: The Case of Mexico." *BioScience* 65(2): 164–73.

Scheffer, M., S. Carpenter, J. A. Foley, C. Folke, and B. Walker. 2001. "Catastrophic Shifts in Ecosystems." *Nature* 413(6856): 591.

Scheffer, M., S. R. Carpenter, T. M. Lenton, J. Bascompte, W. Brock, V. Dakos . . . and M. Pascual. 2012. "Anticipating Critical Transitions." *Science* 338(6105): 344–48.

Schlager, E., and E. Ostrom. 1992. "Property-Rights Regimes and Natural Resources: A Conceptual Analysis." *Land Economics* 68(3): 249–62.

Segura-Warnholtz, G. 2014. "Quince años de políticas públicas para la acción colectiva en comunidades forestales." *Revista Mexicana de Sociología* 76 (September): 105–35.

SARH. 1979. *Programa Nacional de Desarrollo Forestal*. Mexico City: Secretaría de Agricultura y Recursos Hidráulicos.

SEMARNAT/CONAFOR 2006. *Diagnóstico del comercio internacional forestal de México: 1era etapa de la estrategia forestal mexicana*. Mexico City: Unidad de Cooperación y Financiamiento, Dirección de Comercio Internacional, Secretaría de Medio Ambiente y Recursos Naturales. Accessed August 2, 2016. http://www.cnf.gob.mx:8090/snif/portal/economica/diagnostico-del-comercio-internacional-forestal.

SEMARNAT/CONANP. 2016. "Prontuario estadístico y geográfico de las áreas naturales protegidas de México." 1st ed. Mexico City. https://www.gob.mx/conanp/acciones-y-programas/prontuario-estadistico-y-geografico-de-las-areas-naturales-protegidas-de-mexico?idiom=es.

Seymour, F., and J. Busch. 2016. *Why Forests? Why Now? The Science, Economics and Politics of Tropical Forests and Climate Change*. Washington, D.C.: Center for Global Development.

Shapiro-Garza, E. 2013. "Contesting the Market-Based Nature of Mexico's National Payments for Ecosystem Services Programs: Four Sites of Articulation and Hybridization." *Geoforum* 46 (May): 5–15.

Sierra-Huelsz, J. A., K. A. Kainer, E. Keys, and S. S. Colli-Balam. 2017. "Three Stories Under the Same Hut: Market Preferences and Forest Governance Drive the Evolution of Tourism Construction Materials." *Forest Policy and Economics* 78 (May): 151–61.

Silva, E. 1994. "Thinking Politically About Sustainable Development in the Tropical Forests of Latin America." *Development and Change* 25(4): 697–721.

Silva Herzog, J. 1959. *El agrarismo mexicano y la reforma agraria: Exposición y crítica.* Mexico City: Fondo de Cultura Económica.

Simonian, L. 1995. *Defending the Land of the Jaguar: A History of Conservation in Mexico.* Austin: University of Texas Press.

Simpson, E. N. 1937. *The Ejido: Mexico's Way Out.* Chapel Hill: University of North Carolina Press.

Sims, K. R., J. M. Alix-Garcia, E. Shapiro-Garza, L. R. Fine, V. C. Radeloff, G. Aronson, . . . and P. Yáñez-Pagans. 2014. "Improving Environmental and Social Targeting Through Adaptive Management in Mexico's Payments for Hydrological Services Program." *Conservation Biology* 28(5): 1151–59.

Smith, D. M., B. C. Larson, M. J. Kelty, and P. M. S. Ashton. 1997. *The Practice of Silviculture: Applied Forest Ecology.* 9th ed. John Wiley and Sons.

Skutsch, M., M. Olguín, P. Gerez, C. Muench, G. Chapela, R. Benet . . . and R. Galindo. 2018. "Increasing Inequalities in Access to Forests and Forest Benefits in Mexico." *Journal of Latin American Geography* 17(1): 248–52.

Skutsch, M., C. Simon, A. Velazquez, and J. C. Fernández. 2013. "Rights to Carbon and Payments for Services Rendered Under REDD+: Options for the Case of Mexico." *Global Environmental Change* 23(4): 813–25.

Snook, L. K. 1998. "Sustaining Harvests of Mahogany (*Swietenia macrophylla King*) from Mexico's Yucatan Forests: Past, Present and Future." In *Timber, Tourists and Temples: Conservation and Development in the Maya Forest of Belize, Guatemala and Mexico,* edited by R. Primack, D. B. Bray, H. Galleti, and I. Ponciano, 61–80. Washington, D.C.: Island Press.

Socolar, J. B., J. J. Gilroy, W. E. Kunin, and D. P. Edwards. 2016. "How Should Beta-Diversity Inform Biodiversity Conservation?" *Trends in Ecology and Evolution* 31(1): 67–80.

Solorzano, C. R., and F. Fleischman. 2018. "Institutional Legacies Explain the Comparative Efficacy of Protected Areas: Evidence from the Calakmul and Maya Biosphere Reserves of Mexico and Guatemala." *Global Environmental Change* 50 (May): 278–88.

Sosa Pérez, F., and J. P. Robson 2019. "Migration, Community and Land Use in San Juan Evangelista Analco." In *Communities Surviving Migration: Village Governance, Environment and Cultural Survival in Indigenous Mexico,* edited by J. P. Robson, D. Klooster, and J. Hernández-Díaz, n.p. Routledge: New York.

Stevens, C., R. Winterbottom, J. Springer, and K. Reytar. 2014. *Securing Rights, Combating Climate Change: How Strengthening Community Forest Rights Mitigates Climate Change.* Washington, D.C.: World Resources Institute. www.wri.org/securing-rights.

Stoian, D., D. Rodas, M. Butler, I. Monterroso, and B. Hodgdon. 2018. "Forest Concessions in Petén, Guatemala: A Systematic Analysis of the Socioeconomic Performance of Community Enterprises in the Maya Biosphere Reserve." CIFOR. http://www.cifor.org/publications/pdf_files/brief/7163-brief.pdf.

Szekely, E. Miguel, and Sergio Madrid. 1990. "La apropiación comunitaria de recursos naturales: Un caso de La Sierra de Juárez, Oaxaca." In *Recursos naturales, técnica y cultura: Estudios y experiencias para un desarrollo alternativo*, edited by Enrique Leff, 387–409. Mexico City: Universidad Nacional Autónoma de México.

Tainter, J. 2007. "Scale and Dependency in World-Systems: Local Societies in Convergent Evolution." In *Rethinking Environmental History: World-System History and Global Environmental Change*, edited by A. Hornborg, J. R. McNeill, and J. Martinez-Alier, 361–78. Lanham, Md.: Alta Mira Press.

Tanaka, H. 2012. "Competitive Strategies in Community Forestry: Leveraging Through Second-Tier Collective Enterprises in Mexico and Guatemala." PhD diss., University of Idaho, Moscow.

Tardanico, R. 1980. "Revolutionary Nationalism and State Building in Mexico, 1917–24." *Politics and Society* 10(1): 59–86.

Tarko, V. 2017. *Elinor Ostrom: An Intellectual Biography*. London: Rowan and Littlefield International.

Taylor, P. L. 2000. "Producing More with Less? Community Forestry in Durango, Mexico in an Era of Trade Liberalization." *Rural Sociology* 65(2): 253–74.

Taylor, P. L. 2001. "Community Forestry as Embedded Process: Two Cases from Durango and Quintana Roo, Mexico." *International Journal of Sociology of Agriculture and Food* 9(1): 59–81.

Taylor, P. L. 2005. "New Organizational Strategies in Community Forestry in Durango, Mexico." In *The Community Forests of Mexico: Managing for Sustainable Landscapes*, edited by David Barton Bray, Leticia Merino-Pérez, and Deborah Barry, 125–50. Austin: University of Texas Press.

Taylor, P. L. 2010. "Conservation, Community, and Culture? New Organizational Challenges of Community Forest Concessions in the Maya Biosphere Reserve of Guatemala." *Journal of Rural Studies* 26(2): 173–84.

Taylor, P. L., and C. Zabin. 2000. "Neoliberal Reform and Sustainable Forest Management in Quintana Roo, Mexico: Rethinking the Institutional Framework of the Forestry Pilot Plan." *Agriculture and Human Values* 17(2): 141–56.

Taylor, W. B. 1972. *Landlord and Peasant in Colonial Oaxaca*. Palo Alto, Calif.: Stanford University Press.

Tellez Kuenzler, L. 1994. *La modernización del sector agropecuario y forestal*. Mexico City: Fondo de Cultura Económica.

Tobler, M. W., R. G. Anleu, S. E. Carrillo-Percastegui, G. P. Santizo, J. Polisar, A. Z. Hartley, and I. Goldstein. 2018. "Do Responsibly Managed Logging Concessions

Adequately Protect Jaguars and Other Large and Medium-Sized Mammals? Two Case Studies from Guatemala and Peru." *Biological Conservation* 220 (April): 245–53.

Toledo, V. M. 2010. *La biodiversidad de México: Inventarios, manejos, usos, informática, conservación e importancia cultural.* Mexico City: Fondo de Cultura Económica–CONACULTA.

Torres Rojo, J. M., and J. Amador Callejas. 2015a. "Características de los núcleos agrarios forestales en México." In *Desarrollo forestal comunitario*, edited by J. M. Torres Rojo, 15–38. Mexico City: Centro de Investigación y Docencia Económica.

Torres Rojo, J. M., and J. Amador Callejas. 2015b. "La importancia de los apoyos destinados a promover el desarrollo forestal comunitario en el desempeño de las empresas forestales comunitarios y las comunidades forestales." In *Desarrollo forestal comunitario*, edited by J. M. Torres Rojo, 175–224. Mexico City: Centro de Investigación y Docencia Económica.

Torres Rojo, J. M., A. Guevara-Sanginés, and D. B. Bray. 2005. "The Managerial Economics of Sustainable Community Forestry in Mexico: A Case Study of El Balcón, Técpan, Guerrero." In *The Community Forests of Mexico: Managing for Sustainable Landscapes*, edited by D. B. Bray, L. Merino-Pérez, and D. Barry, 273–304. Austin: University of Texas Press.

Torres Rojo, J. M., and S. Graf Montero. 2015. "Desarrollo forestal comunitario: Lecciones aprendidas, tendencias y perspectivas." In *Desarrollo forestal comunitario*, edited by J. M. Torres Rojo, 269–99. Mexico City: Centro de Investigación y Docencia Económica.

Torres-Rojo, J. M., R. Moreno-Sánchez, and J. Amador-Callejas. 2019. "Effect of Capacity Building in Alleviating Poverty and Improving Forest Conservation in the Communal Forests of Mexico." *World Development* 121 (September): 108–22.

Torres-Rojo, J. M., R. Moreno-Sánchez, and M. A. Mendoza-Briseño. 2016. "Sustainable Forest Management in Mexico." *Current Forestry Reports* 2(2): 93–105.

Trejo, I., and Dirzo, R. 2000. "Deforestation of Seasonally Dry Tropical Forest: A National and Local Analysis in Mexico." *Biological Conservation* 94(2): 133–42.

Trench, T., A. M. Larson, A. L. Amico, and R. Ashwin. 2017. "Analyzing Multilevel Governance in Mexico: Lessons for REDD+ from a Study of Land-Use Change and Benefit Sharing in Chiapas and Yucatán." *CIFOR Working Paper* 236. https://www.cifor.org/library/6798/.

Tucker, C. M. 1999. "Private Versus Common Property Forests: Forest Conditions and Tenure in a Honduran Community." *Human Ecology* 27(2): 201–30.

Tucker, C. M., H. Eakin, and E. J. Castellanos. 2010. "Perceptions of Risk and Adaptation: Coffee Producers, Market Shocks, and Extreme Weather in Central America and Mexico." *Global Environmental Change* 20(1): 23–32.

Tucker, C. M., and E. Ostrom. 2005. "Multidisciplinary Research Relating Institutions and Forest Transformation." In *Seeing the Forest for the Trees: Human-Environment Interactions in Forest Ecosystems*, edited by E. F. Moran and E. Ostrom, 81–104. Cambridge, Mass.: MIT Press.

Tucker, C. M., J. C. Randolph, and E. J. Castellanos. 2007. "Institutions, Biophysical Factors and History: An Integrative Analysis of Private and Common Property Forests in Guatemala and Honduras." *Human Ecology* 35(3): 259–74.

Tyrell, M. L., J. Ross, and M. Kelty. 2012. "Carbon Dynamics in the Temperate Forest." In *Managing Forest Carbon in a Changing Climate*, edited by M. A. Ashton, M. L. Tyrrel, D. Spalding, and B. Gentry, 77–108. New York: Springer.

UCEFO. 1989. *La voz de la Unión campesina.* Unión de Comunidades Forestales del Estado de Oaxaca. Newsletter. May–June.

Ureña Argaez, J. L. 2017. "La cooperativa como modelo de explotación de la madera en Quintana Roo durante el cardenismo." Thesis for teacher of sciences applied to regional studies, Universidad de Quintana Roo. http://risisbi.uqroo.mx/bitstream/handle/20.500.12249/311/HD2951.V73.2017-2628.pdf?sequence=1.

Ureste, M. 2016. "Defender el bosque: Una tarea que en el Estado de México cuesta la vida o la libertad." *Animal Político*, March 14. http://www.animalpolitico.com/2016/03/la-familia-zamora-cuando-luchar-contra-los-talamontes-te-cuesta-la-vida-o-la-libertad/.

Vaca, R. A., D. J. Golicher, L. Cayuela, J. Hewson, and M. Steininger. 2012. "Evidence of Incipient Forest Transition in Southern Mexico." *PLoS One* 7(8): e42309.

Van Vleet, E., D. B. Bray, and E. Durán. 2016. "Knowing but Not Knowing: Systematic Conservation Planning and Community Conservation in the Sierra Norte of Oaxaca, Mexico." *Land Use Policy* 59 (December 31): 504–15.

Vatant, F. 1990. *La explotación forestal y la producción doméstica tarahumara: Un estudio de caso. Cusárare, 1976–1976.* Mexico City: Instituto Nacional de Antropología e Historia.

Vázquez León, Luis. 1992. *Ser indio otra vez: La purepechización de los tarascos serranos.* 1st ed. Mexico City: Consejo Nacional para la Cultura y las Artes.

Vega, D. C., and R. J. Keenan. 2014. "Transaction Cost Theory of the Firm and Community Forestry Enterprises." *Forest Policy and Economics* 42 (May): 1–7.

Velasco Murguía, A., E. Durán Medina, and D. B. Bray. 2014. "Dinámica de la cobertura arbolada en comunidades indígenas con y sin iniciativas de conservación, en Oaxaca, México." *Investigaciones Geográficas Boletín del Instituto de Geografía* 83 (April): 55–73.

Velázquez, Alejandro, J. F. Mas, J. R. Díaz, R. Mayorga, García C. Alcántara, R. Castro, T. Fernández, and J. L. Palacio. 2000. "Patrones de cambio de uso del suelo y tasas de deforestación en México." *Gaceta Ecológica de la Instituto Nacional de Ecología* 62:21–37.

Velázquez, A., E. Durán, I. Ramírez, J. F. Mas, G. Bocco, G. Ramírez, and J. L. Palacio. 2003. "Land Use-Cover Change Processes in Highly Biodiverse Areas: The Case of Oaxaca, Mexico." *Global Environmental Change* 13(3): 175–84.

Vidal, O., J. López-García, and E. Rendón-Salinas. 2014. "Trends in Deforestation and Forest Degradation After a Decade of Monitoring in the Monarch Butterfly Biosphere Reserve in Mexico." *Conservation Biology* 28(1): 177–86.

Vidal, O., and E. Rendón-Salinas. 2014. "Dynamics and Trends of Overwintering Colonies of the Monarch Butterfly in Mexico." *Biological Conservation* 180 (December): 165–75.

Villavicencio Valdez, G. V., E. N. Hansen, and J. Bliss. 2012. "Factors Impacting Market-place Success of Community Forest Enterprises: The Case of TIP Muebles, Oaxaca, Mexico." *Small-Scale Forestry* 11(3): 339–63.

Wakild, E. 2011. *Revolutionary Parks: Conservation, Social Justice, and Mexico's National Parks, 1910–40.* Tucson: University of Arizona Press.

Walker, B., C. S. Holling, S. R. Carpenter, and A. Kinzig. 2004. "Resilience, Adaptability and Transformability in Social-Ecological Systems." *Ecology and Society* 9(2): 5. http://www.ecologyandsociety.org/vol9/iss2/art5/.

Wall, D. 2014. *The Sustainable Economics of Elinor Ostrom: Commons, Contestation and Craft.* New York: Routledge.

Wexler, Matthew B. 1995. *Learning the Forest Again: Building Organizational Capacity for the Management of Common Property Resources in Guerrero, Mexico.* PhD diss., Boston University.

Williams, M. 2007. "The Role of Deforestation and World-System Integration." In *Rethinking Environmental History: World-System History and Global Environmental Change,* edited by A. Hornborg, J. R. McNeill, and J. Martínez-Alier, 101–22. Lanham, Md.: AltaMira Press.

Williamson, O. E., 1975. "Markets and Hierarchies: Analysis and Antitrust Implications: A Study in the Economics of Internal Organization." University of Illinois at Urbana-Champaign's Academy for Entrepreneurial Leadership Historical Research Reference in Entrepreneurship. https://ssrn.com/abstract=1496220.

Wilshusen, P. R. 2005. "Adaptation or Collective Breakdown? The Emergence of 'Work Groups' in Two Forestry *Ejidos* in Quintana Roo, Mexico." In *The Community Forests of Mexico: Managing for Sustainable Landscapes,* edited by David Barton Bray, Leticia Merino-Pérez, and Deborah Barry, 151–82. Austin: University of Texas Press.

Wilshusen, P. R. 2009. "Social Process as Everyday Practice: The Micro Politics of Community-Based Conservation and Development in Southeastern Mexico." *Policy Sciences* 42(2): 137–62.

Wilshusen, P. R. 2010. "The Receiving End of Reform: Everyday Responses to Neoliberalisation in Southeastern Mexico." *Antipode* 42(3): 767–99.

Wilson, P. N., and G. D. Thompson. 1993. "Common Property and Uncertainty: Compensating Coalitions by Mexico's Pastoral 'Ejidatarios.'" *Economic Development and Cultural Change* 41(2): 299–318.

World Bank. 1995. *Mexico: Resource Conservation and Forest Sector Review.* Washington, D.C.: World Bank, SARH. http://documents.worldbank.org/curated/en/998301468757196117/Mexico-Resource-conservation-and-forest-sector-review.

"Welcome to the FSC Public Search." 2020. FSC. https://ic.fsc.org/en/choosing-fsc/certificate-database.

Wunder, S. 2005. "Payments for Environmental Services: Some Nuts and Bolts." CIFOR Occasional Paper No. 42. https://www.cifor.org/publications/pdf_files/OccPapers/OP-42.pdf.

Yannakakis, Y. 2008. *The Art of Being In-Between: Native Intermediaries, Indian Identity, and Local Rule in Colonial Oaxaca.* Durham, N.C.: Duke University Press.

Young, O. R., F. Berkhout, G. C. Gallopin, M. A. Janssen, E. Ostrom, and S. Van der Leeuw. 2006. "The Globalization of Socio-Ecological Systems: An Agenda for Scientific Research." *Global Environmental Change* 16(3): 304–16.

Zabludovsky, K. 2012. "Reclaiming the Forests and the Right to Feel Safe." August 2, *New York Times*. http://www.nytimes.com/2012/08/03/world/americas/in-mexico -reclaiming-the-forests-and-the-right-to-feel-safe.html?_r=0.

Zarzosa L., Oscar. 1958. "Problemas de la industria forestal de Durango." *Mensajero Forestal* 16 (162): 273–83.

Zúñiga, I., and P. Deschamps. 2013. "Política y subsidios forestales en México." Consejo Civil Mexicano para la Silvicultura Sostenible, A.C. https://www.ccmss.org.mx/wp -content/uploads/CCMSS_Subsidios_Forestales_190513.pdf.

INDEX

facilitative political regime(s), 14, 17, 20, 36, 71, 99, 125, 224

five capitals, 4, 18, 20, 22–23, 25, 33, 34, 43, 71, 86, 124, 223–24, 227, 230, 237, 246, 248. *See also individual capitals* under capital

FONAFE (National Fund for Ejido Promotion; Fondo Nacional de Fomento Ejidal), xii, 54, 56–59 passim, 60–61, 67, 68, 69, 106, 113, 132, 176, 173, 221

Forest Associations (FAs), 58, 66, 67, 68, 83, 152, 170, 173, 178, 176, 178, 179, 180, 184, 245

forest cooperatives, 27, 42, 44–46 passim, 48, 198, 220

Forest Law: 1926, 42, 198, 219; 1942, 46, 199; 1964, 199; 1986, 36, 68, 73, 74, 114, 200, 222; 1992, 75, 78, 94, 95; 1997, 79, 94, 95; 2003 (General Law of Sustainable Forest Development—LGDFS), 80, 92, 129, 180, 189, 190, 216; 2018, 85; enforcement of, 79

Forest Management Programs (FMPs), 80, 94, 95, 98, 114, 128-130, 140, 164, 169, 170, 174, 175, 186, 189-191 passim, 196, 217

forests, biodiversity of, 4, 5, 6, 35, 37, 38, 39, 81, 82, 185, 188, 197, 200, 201, 204, 212, 213, 215–17 passim, 234, 237, 238, 240, 246, 247; pine-oak, 37, 38, 39, 185, 188, 189, 193, 196, 201, 213, 216; temperate, 4, 37, 39, 87–89, 159, 163, 186–88, 192–94, 196–97, 198, 202, 230, 236, 241; tropical, xiii, 3, 4, 37–40, 56, 57, 85, 87, 88–89, 95, 96, 97, 130, 141, 158, 159–60, 163, 170, 174, 186–88, 190, 192–94, 196, 197, 202, 203, 204, 207, 213, 214, 215, 236, 238, 241, 246. *See also* biodiversity, conservation of

Forest Stewardship Council (FSC) certification, 10, 83, 116, 118, 135, 149, 182, 208–10 passim, 233; barriers to, 208–9; environmental benefits of, 210

FOVIGRO (Vicente Guerrero Forest Company; Forestal Vicente Guerrero), xii, 54–55, 112–14, 119, 120

Fox, Vicente, 79–82, 91

Guerrero, 27, 28, 37–39 passim, 50, 53–55 passim, 58, 59, 75, 78, 80, 87, 93, 102, 103, 111, 112, 117, 118, 121, 146, 148, 149, 173, 174, 195, 216, 220, 221

Hardin, Garrett, 12, 13

ICICO (Indigenous and Peasant Communities of Oaxaca; Integradora de Comunidades Indígenas y Campesinas de Oaxaca), 241

ICOFOSA (Communal Forest Integrator of Oaxaca; Integradora Comunal Forestal de Oaxaca, S.A. de C.V.), 183

IFRI (International Forestry Resources and Institutions Program), xii, 231

illegal logging, 52, 75, 85, 86, 93–95, 99, 117, 147–49, 179, 194

INI (National Indigenous Institute; Instituto Nacional Indigenista), xiii, 48–49, 50, 220

institutions, 4, 7, 14–19 passim, 21, 24, 29, 30, 31, 36, 39, 40, 43, 45, 61, 73, 81, 86, 102, 104, 106, 108, 120, 122, 125, 127, 131, 141, 142, 147, 162, 183, 188, 227, 229, 231, 232, 234, 248; common property, 7, 39, 43, 228, 244; multilevel governance and, 8, 14, 233; and organizations, 17, 18, 20, 23, 33, 49, 71, 74, 100, 121, 122, 124–28 passim, 150, 151, 152, 226, 237, 248

institutional supply, 14, 19, 25, 34, 36, 37, 39, 43

Inter-American Foundation (IAF), 5, 67, 76

Ixtlán de Juárez, xv, 57, 75, 145, 167, 182, 183, 190, 216, 258, 260

Jalisco, 60, 78, 79, 80, 92, 93, 95, 240

Joint Forest Management (India), xiii, 234

LGEEPA (General Law of Ecological Equilibrium and Environmental Protection; Ley General del Equilibrio Ecológico y la Protección al Ambiente), xiii, 214

ABOUT THE AUTHOR

David Barton Bray is a professor in the Earth and Environment Department at Florida International University. He received his PhD from Brown University in 1983. From 1986 to 1997 he was foundation representative with the Inter-American Foundation, and prior to that he was at Tulane University from 1983 to 1986. He has received research funding from the Fulbright Program, the Ford Foundation, the Hewlett Foundation, the Tinker Foundation, and the U.S. Agency for International Development. He is the lead editor of the book *The Community Forests of Mexico* and is widely published in academic journals and in popular outlets such as the *New York Times* and the *Miami Herald*.